Sarah Stuart Robbins

Butterfly's flights : Montreal

Sarah Stuart Robbins

Butterfly's flights : Montreal

ISBN/EAN: 9783337270124

Printed in Europe, USA, Canada, Australia, Japan

Cover: Foto ©berggeist007 / pixelio.de

More available books at **www.hansebooks.com**

BUTTERFLY'S

FLIGHTS.

BY

THE AUTHOR OF

THE "WIN AND WEAR" SERIES.

———◆———

MONTREAL.

———◆———

SAINT PAUL

D. D. MERRILL COMPANY

CONTENTS.

PAGE

I.—MR. GLIDDON'S STORIES 5

II.—WHAT GUY AND BUTTERFLY SAW. 26

III.—BUTTERFLY AT SEA 46

IV.—SAILING THROUGH LAKE ONTARIO. 66

V.—NIGHT ON BOARD THE BOAT . . 91

VI.—THE WOLF STORY 109

VII.—RIVER ST. LAWRENCE 127

VIII.—GOING THROUGH THE RAPIDS . . 147

IX.—COMING INTO PORT 167

X.—LETTERS 187

XI.—HAL ACTS AS CHAPERON . . . 202

XII.—GOOD-BY 223

Order of the Volumes.

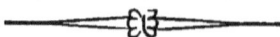

1.—Butterfly at Mount Mansfield.

2.—Butterfly at Saratoga.

3.—Butterfly at Niagara.

4.—Butterfly's Trip to Montreal.

5.—Butterfly at the Sea-side.

6.—Butterfly in Philadelphia.

TRIP TO MONTREAL.

I.

MR. GLIDDON'S STORIES.

ROM Niagara to Lewiston is only seven miles. Butterfly and Guy had hardly time to think themselves in the cars, and realize that they had commenced a trip of four hundred and forty miles, when the train began to slacken, Aunt Matilda to pick up her bundle, Aunt Bessie to tie her bonnet-strings, Guy's mother to say "Guy!" which meant, as the boy well knew, "Be attentive and on the alert now, my son," and there they were in Lewiston.

You may think I do not go very straight forward, when, after telling you this, I begin to give you the reason why the car-ride had been so short; but, as this commences a new volume, I do so to be sure you remember who make our travelling party. The reason was, because the English lady and gentleman, Mr. and Mrs. Gliddon, were on board, and as soon as the farewells were spoken between Guy, Butterfly, and their little friends, Mr. Gliddon had called the children to him and said :

"I have been story-gathering since I saw you. I want you to hear what I have found. Come now, and sit close by me, while I talk. My voice is pretty strong, but now and then the iron horse can neigh louder."

Guy and Butterfly, always interested when a story was to be told, were not long in taking their seats just in front of Mr. Gliddon; the conductor, an obliging man, very kindly turning the seat over for them.

" Once upon a time," said Mr. Gliddon, smiling.

" That is just the way we children begin," interrupted Guy. " How came you to think of it?"

" Oh, I was a boy once, though you might not think it," said the big man. " And I can remember how nice it was to have a little uncertainty about the time of my story; besides, I have another reason. What I am going to tell happened after there had been a war between two countries that were cousins, if not a little nearer."

" Oh, I understand," said Guy. " Cousin John and Cousin Jonathan."

" Who are they?" asked Butterfly.

" Why, don't you know? I thought everybody did." And Guy looked at Butterfly in much amazement.

" No, I do not," said truthful Butterfly, blushing, but not in the least inclined to equivocate.

" Then I must tell you. Cousin John is John Bull, *alias* England, and Cousin Jonathan is the great United States of America."

" Little England and great America. That will do to begin with. Now let us go on. Once upon a time there was a misunderstanding, and they foolishly tried to settle the difficulty by going to war. The battles were fought mainly on the water; and as the large lakes lie between the States and Canada, there were ships brought on to these lakes, and fighting done here. But that has nothing further to do with my story than to explain how it happened that there were three old vessels on Lake Erie, large British ships, declared unfit for service and condemned."

"What does that mean?" asked Butterfly.

"Oh, unfit for service—that tells you plain enough. Leaked, perhaps, or there are a hundred things that might have happened; but it is not often that a vessel condemned gets turned out of use," said Guy.

"These did, though," went on the

story-teller, " as you shall hear. When it was known that they were condemned, there came a number of petitions to the officers to allow them to be sent over the falls. You must remember they were very large vessels, and people thought it would be a splendid sight to see them go pitching over and over, down into that awful abyss."

" I should like to have been there. Did they let them go ?" asked Guy.

" Yes, the authorities had no objection. So, on the appointed day, a great crowd collected along the banks of the river, near the falls, and the vessels were got into the current and left to their fate. In order to add to the interest of the scene, some one offered ten dollars for the largest piece of wood that should be found after the vessels had gone over the falls, five dollars for the second, and so on; and as some of the spectators were foolish enough to suppose a vessel might go over and be seen afterwards, there were many who were on the alert to earn the money."

"Foolish fellows, I am sure," said Guy.

"Yes, yes. You shall hear. Away went the three vessels in the current. They had some old torn flags flying from their mast-heads, and many gay streamers, floating away as merrily as if they had started on a holiday excursion, the poor old ships!"

"I am sorry for them," said Butterfly, the tears dimming her blue eyes just a little.

"That is foolish, too," said stanch

Guy. "They were only wood, without anything to hurt—no life, or soul, or feel-ing ; yet, if they were brave old ships, I am sort of sorry."

"They were brave, of course," said Mr. Gliddon, innocently, "if they had the British flag above them ; but it was better to die a gallant death like that than to come to pieces under the hammer. I think, if they had souls, they would have preferred it."

"So do I, if they must die," said Guy. "Go on, please. What happened next?"

"For miles they kept on pretty well together. Then they got into the rapids, and the oldest and weakest was torn to shivers, and went over in fragments. The next rode on up in sight of the falls, then filled with water, and went down ; but the third, the noblest of the three, took the leap gallantly, and was seen retaining her form perfectly until she was lost in the mists and foam below."

"Bravo for the brave old ship!" said Guy, clapping his hands. "Did they ever find her?"

"No; only one piece of all these three large vessels was ever seen. That was about a foot long, was mashed as by a vice, and its edges were notched like the teeth of a saw."

"O my!" said Butterfly, holding her breath.

"You may well say 'O my!'" said Mr. Gliddon, putting his hand affectionately upon her. "My story shows you the mighty power of these falls; and"—sinking his voice a little lower—"it tells you, too, how great the God is who made them, and how very, very good to love and care for such little, insignificant creatures as we are."

The children both seemed struck by this thought; and, after a moment's silence, Guy said:

"But, Mr. Gliddon, my father would say that mighty water, grand as it is, is a very small thing in comparison with the immortal soul God has given us; and, after all, we are greater than even these great falls."

"And your father would be right," said Mr. Gliddon, looking with much interest at the child, whose education among older people gave him now and then an unnatural maturity of thought and expression.

"Worth more than sparrows—even such little pickaninnies as you," he said, with a genial smile, which quite restored Butterfly to herself. What with the big vessels tumbling over the falls, Mr. Gliddon's unexpected moral, and Guy's sage remark, she was becoming a little bewildered.

"I can tell you another story, about another ship that was sent over the falls."

"Oh, do, do!" said both children.

"Well, there was such a desire felt to see the thing done again, that, some years later, a few men purchased a large schooner."

"How large?" asked Guy.

"About one hundred and forty tons burden."

Guy nodded, as if he underst
fectly just how large that was.

"And," went on the story-teller, "had
it towed down the river to within half a
mile of the rapids, when it was cut adri.
and left to its fate. By the way, do yo
know what makes these rapids?"

"Rocks, I suppose," said Guy.

"Yes, rocks, from two to four feet high,
extending wholly across the river, over
which the water pitches successively
for about a mile immediately above the
main cataract."

"I did not know they went clear
across," said Guy.

"Yes, they do here. And the vessel
took the first ledge bravely, but when
she came to the second, as she pitched
down, over went her masts. And now

came the most interesting part of my story. Those who have lived longest by the falls, and know most about them, say nothing ever went over them and came up alive, not even fish or water-fowl. Numbers of dead fish are seen every day in the gulf below the falls, which are supposed to have tumbled over, and now and then a wild fowl is found that had fallen asleep above the rapids, and so floated on to its death."

" Poor thing!" said Butterfly.

"Would they avoid it awake?" asked Guy.

" Almost always. The instinct of self-preservation is so great that a fowl would rarely swim to its death. But to set the point at rest, the men who had purchased this schooner put two bears and some

other animals on board, and they went
on quietly until the schooner had taken
the second ledge. Then she sprang a
leak and began to fill with water ; and
no sooner did the bears see the water
come rushing in than, like good, sensible
bears as they were, they seemed to have
a misgiving all was not right, and, walking
about on deck, were seen to be looking
for some means of escape. After spring-
ing the leak, the vessel turned stern
foremost and floated along more quietly.
So what should these wise bears do but
step overboard and swim ashore."

"Truly, Mr. Gliddon?" asked Guy, in-
credulously.

"I do not suppose there is a doubt of
it. They had a great deal of difficulty,
and they were carried half way down to

the Horse-shoe Fall by the rapidity of the current before they succeeded in reaching the shore. But they did at last.

"Hurrah for the bears!" said Guy, swinging his cap over his head, much to the amusement of the passengers in the cars. But Butterfly asked as usual, wanting to know the end of the story :

"What became of them then?"

"I don't know. Perhaps they were allowed their freedom after that. I think they deserved it."

"Perhaps their fright tamed them, and they were ready when they came ashore to promise to do no harm for all the future years of their lives, like eating up innocent children and pretty lambs," said Guy.

"Perhaps so, no one can tell. And, fortunately for our story, no one knows."

"But about the schooner?"

"Yes. The schooner, after she changed her position to stern foremost, seemed to lose heart, and went swinging along, until she came to the great Horse-shoe Fall. Then she steadied herself for one moment, as if reflecting whether she would leap or not, and over she went, her bowsprit being the last thing that was seen of her."

"But she came up again?" asked Guy.

"Never—though many of her pieces did. They were picked up miles away, in small bits, bruised, torn all to shivers; hardly, it may be said, a whole atom of her left."

"And the animals — the poor, poor helpless things?" asked Butterfly.

"Yes—a little girl should think of them,

None ever came to shore but the bears, as I have told you. If a fish or a water-fowl could not go over the falls, an animal would have very little chance."

"I think it was cruel," said Butterfly, indignantly.

Just at this moment the car-whistle sounded vigorously, and Mr. Gliddon said, much to the children's surprise, "Here we are at Lewiston;" and Guy's mother called, as I have already told my readers, "Guy!" so the children hastily gathered up their things, and by the time the cars were in the station they were ready to get out.

The station was some distance from the steamboat landing, and now there was to come piling into stage-coaches, which was to show the true character of the travellers, for the coaches could not contain all who wished to go. The walk to the boat would be long, the road sandy, the day hot, and the boat off at the appointed time without waiting for passengers. All this Guy ascertained, after running about and making a few inquiries, then he returned to consult with the ladies of his party.

"We cannot walk, that is certain," said Aunt Matilda.

"No, Guy is the only one that can do so," said his mother.

"I can," said Butterfly.

"Yes, she is as good as a boy," said Guy. "Do let her. If we may both walk, perhaps there can room be made for you three ladies. I will see."

In a moment he returned with a radiant

face. " The driver says," he said, " if we
will walk, he will see you over the road in
some way, and no fail. I think we can
trust him, he looks clever. So, Butterfly,
we will be off with those people that are
just starting. There they go—come quick!"

"There is no trouble in their going,"
said Guy's mother, in reply to an inquiring
look from Aunt Bessie, " if you are not
afraid of Butterfly's getting too tired."

" I am not. She is strong, and loves a
walk, and I know we can trust Guy to
take care of her."

Guy placed himself a step nearer Butterfly, and looked up in Aunt Bessie's face
with a pleased, grateful look, but did not
answer her. Then he took hold of Butterfly's hand, and touching his cap to the
ladies, in quite as old and gentlemanly a

way as his father would have done, the two children followed the few who had been obliging and ready to walk.

A gentleman saw the children coming, and stopped for them.

"You are active young travellers," he said, as they approached. "I love to see children who know their feet were made to use."

"We are very fond of walking," said Guy

"Are you brother and sister?"

"O no. My name is Guy Harrington, and this is Butterfly—I mean, Ellen Courtland. We are not relations."

"So, so," said the old gentleman, with a merry twinkle in his eye. "That is beginning early."

Guy did not know what he meant, but Butterfly did. She was sure he meant they were young to be travelling, and walking part of the way too. So she said :

"O no, sir! I am ten years old, and I have my aunts with me, and they like to have me walk whenever I can. They don't love a lazy child. I have heard them say so a hundred times."

"Sensible aunts, those of yours," said the gentleman, with another smile. "I must make their acquaintance. But while you are away from them, let us see all we can. What is that, right before you? Can you tell me? I am quite sure you neither of you ever saw anything as fine before."

Guy and Butterfly both looked in the direction of his finger, and Guy said, "The great bridge!"

II.

WHAT THEY SAW.

S Guy uttered the words, "The great bridge," with which my last chapter closed, Butterfly, dropping his hand, turned and made a few hasty steps back. This then, right before her, was the great Suspension Bridge, which, when it was erected, she had heard Aunt Matilda say was the biggest and most wonderful bridge in the world. Here was the bridge—but where were the aunts?

"Butterfly!" called Guy, in much astonishment, "where are you going?"

"To tell them," answered Butterfly, a little doubtfully.

"Tell who, and what?"

"Why, my aunts, that here is the great bridge."

"As if they had not eyes in their heads to see it for themselves!"

"So they have," said Butterfly, stopping abruptly, and then coming back; "but, O dear! isn't it splendid! Tell me all about it, Guy." And taking hold of his hand again, she looked up in his face eagerly for information.

"All I know about it is very easily told," said Guy, laughing. "There is the bridge. Can you tell us, sir," speaking to the gentleman who had waited for them, "anything more?"

"I think I can," answered the gentleman, pleasantly. "When it was made, it was considered one of the wonders of the world. It connects, as you see, Canada and the United States, and belongs to a company formed from men of both countries."

"Who built it?" asked Guy, always desirous to know who had done a great thing.

"A gentleman of the name of Serrell, from Canada East. You see, it was very desirable to have this way of communicating between the two countries; but to make a bridge here seemed almost impossible. Let me tell you the dimensions of this. Then, if you know anything of figures, you will have some idea how large it is."

"I don't know that we shall remember

them, but we shall be glad to be told," said Guy.

"I hardly think you will. Yet sometimes a difficult thing like this is the very one a child remembers when it forgets a great deal simpler one. I wonder if you know a child's mind, like its body, grows by what it feeds on; so the stronger the food, often the stronger the growth."

"Tough figures will make tough minds, then," said Guy, smiling.

"Something like that. Trying to remember a hard thing will make you able to do hard work; and that is what you must do if you are to make a man one of these days," laying his hand kindly on Guy's shoulder.

"It is what I intend to be, sir," said

Guy. "And now for the pill, or the tonic, or whatever you would call it, while we have the bridge before us to help us swallow it."

"This bridge, you see, has a very long span. It is one thousand and forty-five feet long. Now, of course, such a length must have very firm supports; so there they are, large towers of cut stone!"

"What makes them stand in the water?" asked Guy, looking at them curiously.

"They are secured by anchors sunk firmly into the solid rock; and then, on each side, there are five cables."

"I know about cables. They are big ropes, or wires, or something that holds up, made of ever so many small cords."

"Yes, each cable here is made of two

hundred and fifty strands of wire. Only think of that!"

"Five times two hundred and fifty make twelve hundred and fifty wires," said Guy, readily. "That is a great number."

"Yes, and every one of these wires is twelve hundred and forty-five feet in length. Now multiply the length by the width, and see how much wire there is there."

This was too long a sum for Guy to do in his head, and I am not sure but Butterfly might have made a mistake if she had attempted it upon her slate; so Guy said, after one or two useless efforts, that he could not make it out; and the gentleman confessed that he had been trying also, but could not get beyond the first two figures.

After this confession, the children stood
in much less awe of him, walked nearer
to him, and began to chat in a free-and-
easy way, which pleased the gentleman
very much.

" What is that?" asked Butterfly, stopping quite still, as she often did when anything interested her, and pointing to a tall
stone shaft, on the side of the river opposite Lewiston, where the bridge ended.

" That is Brock's Monument, and the
town it is in is Queenstown."

" And Queenstown is in Canada; therefore it is named for her right royal
majesty Queen Victoria," said Guy.

" I presume so; but it is more noted
for the gallant defence the British made
here in the war of 1812, than for anything
else. There was a brave officer, Sir

Isaac Brock, killed at that time. The English are very fond of putting up monuments to mark any worthy action. If a man dies bravely fighting for his country, they like to build something which shall say to those around, 'See here! England remembers, and wants everybody that comes here to remember also, what her children have done for her!'"

"Mr. Gliddon will like that," said Guy, turning to Butterfly; "for he thinks there is no other country in the world quite equal to old England."

"And Marie says France; and I say the United States of America," answered Butterfly.

"So do I! Three cheers for the stars and stripes!" As Guy said this, he took off his cap and swung it round his head.

But as for that cap of Guy's, it was taken off, tossed up, swung round, and treated in general in such an expressive way that the only wonder was it ever staid on his head at all. Of its own accord, as soon as the ideas got into the busy brain below it, you would almost have expected to see it starting away.

" I wish we could go to this monument," he ended by saying ; " but I suppose,"— with a droll smile—" even if we should ask the captain of our steamboat to wait for us, he would hardly be obliging enough to do so."

" I am afraid not," said the gentleman ; " but as I have been there and you cannot do any better now, suppose you see it through my eyes?"

" I should like that right well, if you

please, sir," said Guy. But Butterfly, not
quite understanding him, wondered how
they could do so, particularly as the
gentleman wore spectacles ; so she looked
at him with such an expression of wonder
on her face that, half understanding her,
he smiled.

"This is the way," he said, answering
her without any further explanation.
"The monument is one hundred and
eighty feet high, but instead of being
plain stone, as it looks from here, it is
made up of a variety of stones. First,
there is a great stone, forty feet square,
with four lions on it."

"Real, live lions?" asked Butterfly.

"O no; stone lions. The emblem of
England is a lion, you know, so they are
put on every patriotic thing."

"As we use the eagle," said Guy.

"Yes, in the same way. Well, above this comes the base of the pedestal, and then the pedestal, and then the base of the shaft, and then the shaft itself, and then a Corinthian capital, on which is wrought a statue of the goddess of War."

"That is good," said Guy, "as Brock died in war."

"Yes, very appropriate. Then on this capital is a dome, and on this dome is a statue of General Brock."

"It must be pretty large, to be seen with all those things under it, I should think," said Guy.

"It is a colossal statue, which, I sup-pose, you know means very large."

Now Butterfly had seen pictures of mon-uments, and had always supposed they

were just tall wooden steeples with a man
or a horse on the top, as they had a
weather-cock on the top of their meeting-
house steeple at home. She was, therefore, very much surprised to hear that it
took so many things to make up a real
one, and she walked along with her eyes
fixed upon this, trying to remember all
she had been told about it, until, hap-
pening to see the bridge again, she was
a little frightened for fear she should
forget the number of cables, wires,
feet, and so forth; and the consequence
was that the whole information began to
tumble out of her memory as fast as it
could. The feet kicking the wires, the
wires letting the cables slip, the cables
pulling down the towers, the towers
pulling up the anchors, and the anchors

dragging down the whole bridge, and then the monument came tumbling after it. Away went General Brock, and the dome, and the Corinthian capital, and the shaft, and the base of the shaft, and the pedestal, and the base of the pedestal; and even the lions rolled off the sides of the sub-base, leaving nothing there but a great square stone, after all. And then there was the water, and the long steamboat, so much larger than the Minnehaha on Lake George, and stage-coaches rattling in, and—yes, there could be no mistake—Aunt Matilda, Aunt Bessie, and Guy's mother so near them, dashing up the road, that she heard Aunt Bessie say, "There are the children, safe!"

And then Guy's cap had to come off, of course, and there was a great deal of waving done, with about as much shouting as if Guy and Butterfly had just been saved from a wreck on Robinson Crusoe's island. But I must say for Butterfly, that her part of this was quiet and ladylike.

There had been some doubt in the travellers' mind whether they would take the American or English steamer to Montreal, but they had decided to take the American, Guy being very desirous to patronize everything American while he could, and the others caring little; so now, without delay, they went on board their steamer, and Butterfly, experienced traveller as she already thought herself, found a new world.

The boat was crowded with passengers, and as the children looked around, there was not a face among them that they had ever seen before.

"I thought Mr. and Mrs. Gliddon were coming," said Butterfly, much disappointed at missing her new, kind friends.

"Catch Mr. Gliddon," answered Guy, "on board an American boat, when there was an English one to be found." And then Guy congratulated himself on his own choice having been adopted. "We men—I mean, men generally," said Guy, correcting himself, "like to stick by our own flag, if we can."

"I like the new things," said Butterfly, innocently.

"So do I, but I always choose American."

This was something Butterfly could not understand. I doubt whether she would have felt just so, if she had; but she is only a little girl, you know, so you must excuse her.

While the children were looking about them, and talking busily, the aunts and Guy's mother had been hunting up their state-rooms, the numbers of which were upon the tickets they had bought in Niagara. As soon as they had found them, Aunt Bessie came for Butterfly.

"Come," she said, "and see what a cunning little room you are to have for the next twenty-four hours."

In her haste to follow, B utterfly forgot to take her travelling-bag. She had put it down close by a seat when she first came on board, and that v as the last she thought of it until Aunt Bessie said, as she turned the key of the state-room:

"We will put our bags in here, and you

can take off your hat, if you wish. This boat is our hotel for the present, and we shall live here as we did at the Clarendon."

But Butterfly heard very little of what Aunt Bessie was saying. Everything else was lost in the thought of the bag.

"O dear! Guy! Aunt Bessie! O dear me!" she said, clasping her hands together, and looking very much frightened.

"What is it?" asked Aunt Bessie, startled.

"My bag—I've lost it!"

Now, as this was the first thing Butterfly had lost since she had been on her travels, Aunt Bessie did not, even when she first heard of it, feel disposed to blame her. So she said, quietly:

"I do not think it is lost. Don't be frightened. Try to recollect where you had it last. I saw it in your hand when you came on board of the boat."

"Yes, ma'am," said Guy, "she put it down on deck. If it has not been stolen, it is there now. Come, Butterfly!"

So the children rushed up on deck—in a way which made timid people that saw them think some accident was about to happen—to the spot where Guy thought she might have left it. But how different everything looked, now they were confused and in a hurry. The cabin door seemed to have turned itself about to where the smoke-pipe was, and the smoke-pipe to have taken its station where the cabin door was. The long row-boats that they knew were swung on the middle of the steamer seemed to have gone to the stern, and the wheel to have followed them. In short, everything had turned topsy-turvy while they had been in the cabin.

"Little miss," said a gruff voice, swinging Butterfly's bag under her very nose at the minute she was ready to cry from disappointment and alarm, "if you are looking for this, here it is. I saw you put it down, and then run away; so I kept my eye on it for you. It isn't the safest way of travelling, even for a child. Is there no one on board who has the care of you?"

"Thank you, sir, I have," said Guy, taking the bag, with a nice bow.

"And so have my aunts," added Butterfly. "I do thank you very much indeed. I thought I had lost it forever, and I was so sorry, for I didn't mean to lose a thing all the time. My mother said it would make me troublesome, if I did. I do thank you a thousand times."

"You are very welcome," said the man, smiling.

How tall and large he was, and how grim he did look, for all his smile! Butterfly took her bag, looking up in a very timid way. Something about him made her think in a moment of the lions she had heard about on the side of General Brock's monument, but she would not have said this, even to Guy, for Butterfly was a grateful little girl.

Aunt Bessie was very glad when the children came running into the state-room with the lost bag. She kissed the little excited face lifted up to hers, for this expression of forgiveness, and then said.

"Now we will put our room in order for the trip."

III.

BUTTERFLY AT SEA.

OOM, Aunt Bessie!" This, after sitting some moments in silence, looking around the state-room, was Butterfly's first exclamation, accompanied by a merry laugh. "Do you call this bit of a cubby, a real, live room?"

Now Aunt Bessie understood that when Butterfly asked if a thing was *live*, she meant a real, true thing, and not a part of a story or a plaything, so she answered:

"As we are to stay here until we reach Ogdensburg, to-morrow, we shall find it is."

"O my! isn't it splendid! Isn't it the most beautiful, perfect, elegant place you ever saw in all your life, Aunt Bessie?"

"Well," said Aunt Bessie, laughing, "I can't say that it is. You use almost too many of those great words for so little a girl, don't you, Butterfly?"

"Yes, ma'am, mother says I do. But O dear! O dear *me!* I never, in all my whole, live-long life, saw, or knew, or heard, or dreamt of such a be—a—u—ti—fu—l room. Why, I don't believe Princess Adelaide, or Princess Alice, or any of Queen Victoria's little girls, have one half so cunning! Are we really and truly going to have this for our own until we get to Ogdensburg?"

"Really and truly," answered Aunt Bessie.

"Then let's fix up." Saying which, Butterfly jumped from the odd-shaped seat upon which she had been sitting, drew aside some white muslin curtains, and peeped behind them. There were two cunning little beds made up with the whitest of sheets and counterpanes, and the tiniest of pillows.

"What is this?" she asked, after looking a minute.

"These are our beds. The name on board a boat for them is *berths*. How do you like them?"

Butterfly answered by jumping into one, and extending herself at full length.

"I sha'n't sleep one wink," she said, at length.

"Why not?" asked Aunt Bessie.

"O dear! It is so nice! I can't bear

to lose a minute from seeing it. I shall stay awake all night long so not to."

"Shall you?" said Aunt Bessie, gravely.

" All night long. I wouldn't lose seeing it for all the world! Was it made a purpose, Aunt Bessie?"

" On purpose for what?"

"Why, for little girls to be so happy in."

"I rather think it was," said good Aunt Bessie.

" Well, this is the goodest world I ever did live in." And as she said this, Butterfly drew herself upon her elbow, and looked delightedly around her. "It is not bigger than Carrie Jones' baby-house, yet they are all here."

They was not very definite, but **Aunt** Bessie understood her to mean everything

4

that was indispensably needed in a sleep-
ing-room ; so she said :

" Yes, all here, on a limited scale, Butterfly."

" I don't know just what you mean by a
limited scale, Aunt Bessie, but I had
rather have it than the biggest parlor at
Saratoga."

" I dare say."

" I'll put my bag here, and your bag
there. My trunk—where is my trunk, Aunt
Bessie ?"

" With the rest of the luggage. I don't
know where it is kept."

" And are not we to have it here?"

" Where would you put it?"

Butterfly looked around the state-room.
With Aunt Bessie and herself both stand-

ing, I doubt whether she could have found room to put down a quart cup.

"So there isn't," she said, after making a very deliberate survey. "I like it, though, better and better."

"I am glad of it," said Aunt Bessie. "Now I will lie down a few minutes, so you shall have room to do your contemplated fixing up."

Butterfly thought first just what Aunt Bessie could mean by *contemplated ;* but, not being able to satisfy herself, and not liking to be troublesome, she resolved to wait until some other time to inquire, and go to work now.

It was amusing to lie still, as Aunt Bessie did, and watch Butterfly. She was an orderly little girl, and had, besides, a tasteful way of arranging. Her mother had early taught her it costs no more trouble, and brings about a much more desirable result, to do whatever she did in the nicest and prettiest way. To lay her clothes smoothly in her drawers; to have her handkerchiefs put squarely in a pile; her hair-ribbons laid smoothly in a little box, the colors that looked best together side by side; and all the other tasteful things which little girls will love to do if they are willing to *take pains.* These are two very important words—" *Take pains.*" I wonder if my young friends who are reading this book know how much they mean?

Butterfly was taking pains as she arranged her own and Aunt Bessie's travelling things, so the good aunt loved to watch her as she rested.

So fully was Butterfly occupied for a half hour that she forgot everybody and everything else, therefore she was startled when she heard a knock at the state-room door.

"What is that?" she asked.

"A knock on the door. Open it."

"Do they knock here? How funny!" And opening the door there stood Guy.

"Are you never coming out?" he asked, a little impatiently. "I have been waiting here as patient as a saint this long time. I thought we should get to Montreal before you made your appearance, and you have lost seeing the last of the bridge and the monument already."

"Why," said Butterfly, a little confused, "we haven't started yet. I've been waiting for the boat to move, then I was

coming. And, O Guy!" throwing her
door wide open and pulling him in, "did
you ever see such a beauty of a room?"

"There are two hundred on board this
very boat as like it as two peas in a pod,"
said Guy, contemptuously.

"Two hundred!" And Butterfly fairly
held her breath in surprise.

"Of course there are. Where did you
suppose all these passengers were to
sleep?"

"I never thought." Which was very
true. Butterfly had been too busy with
other things.

"Where is Aunt Matilda?" she said,
remembering her.

"She is with my mother. They have a
state-room together next to mine. And I,
poor solitary chap that I am, have to take

in a big man for company. By the way,
I believe it is that very gruff fellow that
kept your bag for you, so he is good-
natured, at least."

"Indeed he is. I thank him a thousand times. Now let us go and see
them."

"See who?" Guy could not get used
to Butterfly's indefinite way of saying
things. She seemed to suppose, because
she knew so well what she wanted to do,
everybody else must know also.

"Why, your mother and Aunt Matilda!"

"Come, then; only hurry, or we shall
be in the rapids before we are on
deck."

Aunt Matilda's state-room was at the
opposite extremity of the cabin from the

one occupied by Butterfly, so, as the
children went towards it, Butterfly had a
fine chance to peep in at the many open
doors of the state-rooms.

No one saw the child, with her beam-
ing, pretty face, so full of amused astonishment, but smiled back; therefore, by
the time Guy tapped at his mother's door,
Butterfly hardly knew how to contain her
happiness within proper bounds.

"O my! Aunt Matilda," she said,
tumbling in head first, "there are two
hundred of them! Guy says there are.
And here is yours besides; and they
are play-rooms, after all. Aunt Bessie
thought they were real live ones!"

Aunt Matilda was so accustomed to
Butterfly's random way of speaking that
she easily understood her, so she said:

"It *is* all very new and droll to you, isn't it?"

"It is the beautifulest place I ever saw; and there were ever so many little girls, and such a lot of shawls, and red curtains in some, and white in the others, and white bed-quilts, and"——

"If you are going to give an inventory of the whole two hundred," said Guy, interrupting her, "that will end us for to-day. We may as well give it up for lost."

"What is an inventory?" asked Butterfly, easily diverted.

"A list of articles contained anywhere."

"That is hardly Webster's definition, I think, Guy," said his mother, smiling.

"No; but it will answer. Ready, Butterfly?"

"When the boat starts, and I have kissed Aunt Matilda," said Butterfly, holding up her lips.

"My kiss too," said Guy's mother.

"How girls do love to kiss!" said Guy, looking on.

"So do some boys," said his mother, bending over him.

Guy did not say a word, but Butterfly heard a sound very like a kiss, then he held his hand out to her, and said, "Come!" So they went' back again through the whole length of the cabin, Butterfly, more at home now, nodding and smiling to the little girls in the state-rooms.

"You talk about starting. Let me see you find the bridge and the monument, if you can," said Guy, as he brought two stools and put them in the stern of the boat, where the view was uninterrupted.

Butterfly looked in all directions, but nothing was to be seen of them; and then, to her surprise, she found the boat was actually on its way. She had no idea before, so quietly did it move, that they had left the wharf at Lewiston.

"After all, we are just in time." These were the first words Guy spoke after they had seated themselves.

" That is Fort Niagara right before you. I thought we had lost it."

" Why, it is not a bit like Fort Ticonderoga," said Butterfly, looking at the fort with intense interest.

"No, because this is a fort and that was not. I wish we had some one here to tell us stories of this fort. It has been a

famous place in the wars of the whites
and the Indians, and also those between
the French and English."

Guy gazed around among the passengers who were nearest them, hoping to
find the wished-for story-teller; but no
one who looked as if he cared in the least
for the little boy and girl was to be seen,
so he began himself:

"I can tell you this much about it. It
stands at the mouth of the Niagara river,
where it empties into Lake Ontario. Now,
before you know it, we shall be in the
lake. You see, this is to keep great vessels from sailing up the river, if we ever
should quarrel with Great Britain and go
to war."

Butterfly had little idea what it meant
to go to war, and so the most she thought

of the fort was that it was a very ugly-looking building, not half as pretty as Fort Ticonderoga, that Guy said was just " no fort at all."

" Pretty soon we shall see an English fort," said Guy. " You must remember, Butterfly,"—Butterfly nodded her head by way of promise that she would— "that Lake Ontario and the river St. Lawrence run between two distinct countries, Canada and the United States ; and that, as we sail along, we shall have English towns on one side and American on the other; just as now, there is Fort Niagara, American, on this side, and— there, you will see it very soon—Fort Massasanga on the other. Let us watch and see if the American isn't a great deal the best."

"Yes," said Butterfly, gravely. "But I don't like forts half as well as I do the water. Oh, Guy, do look there, and see how smooth and beautiful it is, all but where the boat has gone, and that looks—looks—looks"—and Butterfly hesitated, trying to think of something it did look like.

"Like the foam of the ocean," said Guy.

"Yes, I guess so. I never saw it, but it does," answered Butterfly, perfectly satisfied.

"I say," Guy called out, suddenly, "did you see it?"

"See what?" asked Butterfly, starting up and opening her eyes very wide.

"Why, that fish. I should think it was as long as my whole arm," extending an arm which to a man might not have seemed so very long.

" No—where ?"

" Oh, it's gone for good. Butterfly, if the steamboat would only stop a little while, and somebody would lend me a line, I could catch fish enough for a picnic dinner."

" How nice that would be !" said Butterfly. " And we would put off to a little island, as the people do on Lake George, and have a grand time."

" Yes, and we would camp out."

" So we could ! And we would build a fire, and make a nice cup of tea for your aunts and my mother."

" And then at night !"

" What would you do, then ?"

The children turned to see a boy, a

little larger than Guy, standing close by his side. The boy was dressed like an old man, Guy thought, and looked so differently from the children with whom he had been used to play, that for an instant he was rather in doubt whether it was a boy or a dwarf that had spoken to him. But the boy raised his hat politely, and said:

"Excuse me—but there is nothing I like so well as camping out. I heard you talking about it, and I spoke without thinking."

"Sit down," said Guy, moving an empty seat near him. "We were saying, if we could stop the boat, and if we had a line, and if we could land on a little island, and if we could build a tent, what we would like to do."

"I live in Montreal," said the boy, sitting down, "and every summer, in the holidays, my father sends me away with my tutor. There he is, don't you see him?" pointing to a short, square-looking young man, dressed in light clothes, with a tall white beaver. "I can tell you, he is a nice young man, and we have nice times together. We fish, and hunt, and camp out. I will show you, if you go on to Montreal. Are you going?"

"Yes," said Guy.

"I am very glad. I go over here every year. My father says I may go to Niagara until I am tired of it—and I think it will be every year as long as I live."

"So would I," said Butterfly.

The boy nodded approval, and then said, "Here we go into Lake Ontario!"

IV.

LAKE ONTARIO.

HEN the boy said, "Here we go into Lake Ontario," Butterfly started to run down into the cabin for Aunt Bessie. That morning she had heard the stories of the vessels being sent over the falls, and she had her thoughts so full of the way they went tumbling down, that, without knowing it, she had an idea, when they passed from Niagara river into Lake Ontario, they were to go over something that would set the boat into a great whirl, and perhaps upset it; so her first thought was to seek Aunt Bessie's protection.

"Where are you going?" said Guy. "Wait and hear about Lake Ontario."

By this time Butterfly had looked around her, and saw nothing but a smooth expanse of water, the lake and the river running together without the least perceptible change.

"I thought," said Butterfly. And then she stopped short.

"Thought what?"

"That something was coming!"

"So there is—or rather it has come. Here we are now in Lake Ontario, and I can tell you a great deal about it. I learned it on purpose last night. The Guide Book says:

"Lake Ontario is a wonderful sheet of fresh water. It is two hundred and thirty-five feet above the level of the sea, one hundred fathoms deep, two hundred miles long, and sixty wide."

"Correct, Mr. Niles would say," said the new-comer, nodding towards his tutor.

"In crossing it, one loses sight of land altogether."

"How nice!" said Butterfly, clapping her hands. "No land anywhere to be seen! Won't it be droll?"

"Yes, if it is pleasant," said the stranger; "but if a storm comes, look out, that is all! I was on board once when we had a terrific blow. Mr. Niles said he had crossed the Atlantic three times, and he never had seen anything like it. I tell you what, sir,"—addressing Guy, who liked to be called sir—"it was one of the fiercest gales that ever swept this lake. Mr. Niles says a storm in one of these land-

locked lakes is a great deal more dangerous than in the open sea. Even the big
men-of-war get pretty well tossed about."

"I hope we sha'n't have any," said
Butterfly, looking frightened.

"Nothing that looks like one, I will
promise you, to-night. Just look up
there. There isn't a cloud in the sky,
and the wind is right for pleasant
weather. My mother has always been a
little timid about me since that time, and
Mr. Niles promised her not to go on to
the lake unless the day was perfect.
Niles is great on weather. He guesses
right nine times out of ten. This morning
he came into my room. 'Hal!' said he,
'everything is promising, and if you like
we will start to-day.' That is one thing
I like about my tutor. He never says

you must and you must not, as so many of
them have done, but it is always, 'Hal, if
you please,' 'Hadn't we better?' or something that makes a fellow feel like a man."

"That is the way my father does," said
Guy.　"I don't believe, if I was one
hundred years old, he would treat me any
different from what he does now."

Hal laughed.　"My father says, 'Ask
Niles,' 'Go to Niles,' 'Do as Niles says.'"

"I never had a tutor," said Guy.　"I
always go to school.　I am an only child,
you see," he added, apologetically, " and
my father says nothing will be easier
than to spoil me."

"I am an only son," said Hal; "but I
have two sisters, one older than I am,
and one as large as she is," pointing to
Butterfly.

"And your home is in Montreal?" asked Guy. "Then you are a subject of Queen Victoria."

"God bless the Queen!" said Hal, raising his odd-shaped hat. "Thank fate, yes."

Guy wanted to say he did not think being a subject of a monarch was much to be thankful for, but he had been too well brought up not to know it is very impolite to express a contrary opinion unless there is a special reason for so doing; so he looked with some curiosity at a boy who was a *subject*, having in his wise little head no idea that he, Guy Harrington, was other than a free man.

"If you have never been through the lake and the St. Lawrence," went on Hal, without thought that he was exciting

Guy's pity, "you have a treat before you ; and if," added the boy, modestly, "you will take me for a guide-book"——

Guy remembered how many times he had been Butterfly's guide-book, and how pleasant he had found it; so he said, generously:

"There couldn't be anything we should prefer. I do so love printed letters coming out of the mouth, instead of off from the white paper." By which speech, I suppose, Guy meant to say he had rather be taught orally than in any other way.

"There is a little girl, as big as you are," looking at Butterfly. "She is a nice child," said Hal. "Her parents live in the same street where we do, in Montreal. Her father is Lord John Herbert. I have been talking with her ever since we started until Niles called me, and said, "Hal, there is a boy you will like to know. Go and get acquainted with him. If you have no objections, I would like to bring her here."

"Of course, do by all means. The more the merrier. Butterfly and I are always making new friends. There were the Nelsons at Niagara, and Bertie. Ned was a good fellow, but I liked Bertie rather the best, and we are to go to school together this winter."

"Who *is* Butterfly?" asked Hal, without taking any notice of Guy's list of acquaintances.

"Why, this is Butterfly," pointing to the child.

"Really and truly?"

"We all call her so."

"My true name," said Butterfly, "is Ellen Courtland, but it's Butterfly too."

"Oh, I understand. That is your pet name."

Butterfly nodded assent, but could not help wondering, the name sounded so familiar to her, why it should strike others as being strange.

"That little girl's name," continued Hal, motioning towards the child he had left, "is Margaret. I like the name, don't you?"

"I like Maggie, or Meg," said Guy.

"But that would never do." And the color came into Hal's face. "They never say so."

"Why not? Don't they love her?"

"O yes, very much. But—but"—hesitating, "we don't do so in Montreal."

"Don't you?" said Guy. "Then I am glad I live in New York, for I like nicknames. Mine is so short. No one can say anything else. Now, if it had been Nebuchadnezzar or Demosthenes"——

"Or Belshazzar, or"—— said Hal.

"Or," interrupted Guy, not very politely, "'By the help of God I have leaped over a wall.' My father says, in good old Puritan times, a boy was baptized by that name. At least so some of the histories say."

The children all laughed, and then Hal went to call Margaret, who was sitting very demurely watching them from another part of the boat.

After a few minutes' conversation with an elderly lady under whose care the child seemed to be, he returned with her, and Guy politely gave her a seat near Butterfly.

There is nothing very formal in children's introductions. Not a half hour had passed since these met for the first time, and I am quite sure any one looking on and listening would have supposed they had been acquainted all their lives.

Margaret was a good specimen of a nice English child. Ruddy, and with beautiful rounded form, every movement ·was full of ease and grace. Strong and perfectly healthy, her complexion looked almost like a painting; and as for her eyes and mouth, as Guy sat by her, seeing how they both smiled, and how much sweetness there was in them, for the first time since he had known Butterfly he thought there could be in the world somebody else almost as beautiful.

This little girl was, like Hal, dressed in

a way which seemed peculiar to these American children. But of course it was only the more pleasing to them.

Margaret was dressed in a suit of thick brown stuff, with a large flat brown hat, and two long drooping brown feathers. Her gloves and boots were of precisely the same shade, and so was the ribbon which tied the ends of her braided hair.

Butterfly had often heard fairies called 'Brownies,' and her first thought when she saw this child was that she must belong to this favorite class of her many dear friends. But there was too much of the little human child in the way Margaret entered into all they said or did for this fancy to stay long ; so before the boat had passed many miles, her hand somehow found its way into the loving clasp of the

English girl's, and she was sure they would be warm friends very soon.

I must say, in the pleasure of being together, they forgot the lake and the things that were to be seen; for the boat had gone past several points of interest before Hal remembered he was the guide-book. Even then he was recalled to it by hearing some one near him say, "There is a light-house."

"Plenty of light-houses here and on the St. Lawrence," said Hal, "and not much more here. You see, the seeing is to be done after we reach that famous river. Sailing through Lake Ontario is like going to sea, only you know you are not on the ocean, for the lake is quiet as now, unless a storm comes, and then, if it does, as I have told you before, look out."

"Where?" said Butterfly, looking round in every direction, and seeing nothing but here and there a distant white sail.

"Oh, I didn't mean look out and see. Only if you went rolling about here, there, and everywhere, as I have done on board, you have to look out sharp where you put your feet."

"My grandmother was wrecked here once," said Margaret.

"Wrecked? How nice!" said Hal. "Please tell us all about it."

"She will. Let us go and ask her," said Margaret. And then the children all ran across the boat to the old lady with whom Margaret was sitting when Hal called her to join them.

"Grandmamma, we want to hear about your shipwreck," said Margaret, as the children gathered close around her.

"And these are your little friends—Americans, I presume," the old lady said, holding out one hand to Guy and one to Butterfly. "How do you do, my dears? Would you like to hear my story?"

Such a kind, beautiful smile came over the sweet old face, that Butterfly at once put both her hands into the one extended towards her; and then, not contented, she lifted it to her mouth and softly kissed it.

The lady smiled again, and, drawing Butterfly close, also kissed her round, red cheek.

"I am glad my Margaret has found a playmate," she said. "She loves company, and I am always happy when the dear child is."

Margaret took the hand which Guy had dropped, looking very lovingly in her grandmother's face, as she said this.

"Two little girls only, now," said the old lady. "Once I had four. I was going to see one of them, though she was a woman and not a little girl then, when I was shipwrecked."

"Do tell us all about it, Lady Herbert," said Hal.

"I was going to Toronto, to see my daughter, Mrs. Somers, as I have already told you," began the old lady. "We had a dark, cloudy afternoon, the wind groaning through the ship's rigging, and the waves rising higher and higher every hour. Our captain was used to the lake, and did not feel any alarm. He said we should have a rough blow, and it would make those who were not used to the sea pretty sick; but the vessel was new strong, and well-manned. So, when it came time to go to bed, he ordered us all down into the cabin, and said he would put out the lights, as there was danger of fire, the ship rolled so.

" Margaret's mother was with me. She was then about as old and as large as Margaret is now, and we came down the little steep stairs of the ship into the cabin, holding on to everything we could find to help us. Margaret was pitched over and knocked about in a way that would have hurt her at any other time, but she didn't mind it then. She saw I was alarmed, and so she tried to comfort me.

" 'Mother,' she said, 'Jesus can be on

these waters as well as on those of the Sea of Galilee, can he not?'

"'Certainly, Margaret.'

"'And isn't he?'

"'Yes.'

"'Then don't let us be afraid. He will take good care of us, if we ask him to.'

"Now, children," said Grandmother Herbert, stopping and looking around, "I want to tell you how much those few words comforted me. I thought Jesus must be very near, if my little daughter could feel so trustful and quiet, and I asked our Saviour to walk on the stormy waters that night, and to save us, lest we perish. Then we went into our berths, and after lying still some time, Margaret put her head down (she was in the upper berth), and said, ' Mother, is Jesus here still?' and I said, ' Yes, my child.'

" Then the wind rose, and the waves broke against the sides of the ship as if great hands were striking it with a giant hammer. There were other passengers on board—I think fifteen besides ourselves — and we were all fastened down into this dark little cabin together. It was like a tomb, children, and it seemed to grow darker and darker as the storm increased. None of the passengers went into their berths but Margaret and I. They were all over the floor, sometimes on one side of the ship and sometimes on the other, as the waves rolled them about. Very few words were spoken. Now and then we heard a groan or perhaps a short prayer;

or, I am very sorry to say, a dreadful oath, wrung out from a man more used to swearing than praying in his hour of mortal peril.

"It so happened that, when Margaret asked me about Jesus, there was a lull in the storm, and the vessel was still. I did not know any one heard her or my answer, until all was over.

"'Then, mother,' she added, 'I need not be afraid.'

"'No, my darling,' I said. 'Be not afraid : it is I.'

"'Is that what Jesus said?'

"'Yes.'

"Then we were all still, and then again came a rush of wind and wave, and my little girl asked once more :

"'Mother, is Jesus here still? He

won't be frightened away, and leave us, will he?'

"'He is here, darling. Nothing can separate us from him.'

"'Yes, mother.'

"'Amen!' said a man's deep voice from somewhere out of the darkness. 'Little girl, I thank you.'

"I never knew how many hours we spent there. It was so dark in the cabin that none of us saw the day dawn; but there came a crash—an awful crash—and in the stillness that followed I heard once more :

"'Has Jesus gone, mother?'

"'No, my darling, no. He is nearer than ever now. Take fast hold of his hand.'

"'Yes, mother. Let me take yours

too.' So, as I tried to find her in the dark, the dead-light was raised, and the day came streaming down into our darkness.

" 'The ship must be cleared,' said a strong, steady voice, speaking through the open door. 'Plenty of time to save you all, if you will come up quietly and orderly. No fainting nor screaming, now. Let the women come first. Order, there! Steady, now.'

" Something in the quiet of the voice told upon us all. We went up without confusion or noise, and were lowered, one by one, into the long-boats, which were already on the water.

" The waves were still high. Our boat rocked about so unsteadily that no one dared to go down into it.

" ' Is Jesus there ?' asked my little girl again.

" ' Yes, darling.'

" ' Then I am not afraid.' And my Margaret threw herself into a sailor's arms who was waiting for her, and was the first one to be safely put into the unsteady boat."

" That was my mother," said Margaret. looking proudly round upon the listening group.

" Then I followed, and the other passengers one by one, until the boat was full. When we had rowed off a short distance from the ship, she rolled over heavily on to her side, and sank slowly.

" ' That is full two hours before the captain expected it,' said a sailor, as the waves closed over her. 'A better craft than the Victoria never sailed the sea. Bad luck to the storm !'

" 'Thank God, we are safe,' said the man at the helm of our boat. 'Row away there, men. Put her ashore at the nearest point.'

" So they did ; and even before the news of the wreck reached Toronto, we were there to tell of our escape."

"Jesus was with you, wasn't he ?" said Butterfly. "Oh, I am so glad you were saved !"

"Jesus never forgets those who trust him," said Grandmamma Herbert, " be they young or old; and sometimes he puts his words and thoughts of comfort into the mouths of his little children, and they become our teachers and comforters. Children should be very pure and good when their Saviour singles them out so to bless and to give a blessing, should they not?"

"My mother was," said Margaret again, more fondly than proudly now.

"There it is!" said Guy, abruptly. The truth is, he liked the story better than the application. Guy was not fond of preaching, and now he was glad of the diversion occasioned by the sounding of the dinner-gong.

"Now for a rush!" he said. "Come, Butterfly. I peeped in, and the cabin is so small it will be a perfect jam. Hurry, and we will go for our friends."

V.

A NIGHT IN THE BOAT.

UTTERFLY had never taken a meal on board a boat before, and she was therefore very much amused and interested in everything she saw. The crowding and pushing of the passengers rushing from the deck in order to insure the best seats and the best of everything, the marked distinction which appeared at once between those who were ladies and gentlemen and those who were not, struck her with surprise, child as she was, and perhaps gave her a better lesson on the importance

(91)

of early acquiring good manners than she could have learned from any other source.

Guy had gone into the cabin before the rest of the party, and, when Butterfly came down, she saw a row of five chairs turned against the table, Guy walking back and forth behind them, as if he was keeping guard.

As soon as he saw the ladies, Guy began to turn back the chairs, and motioned to them, so they went to the part of the table where he was, and he said to Butterfly :

" Sit next to me, do."

Butterfly was so intent on looking that it was as much as she could do to seat herself in any way. She took half of Aunt Bessie's chair, who was on one side of her, and then half of Guy's when she

tried to move out of it, until Guy laughed
and she thought what she was doing.

"O dear, Aunt Bessie!" she said, "I
should think it had dropped right down
out of the fairies' palace."

"What?" asked Aunt Bessie, smiling.

"Why, everything. Do see : there are
peaches, and pears, and grapes, and
apples, and, how funny, all those little
white fans perched up in the goblets, like
big white moths."

"Those are the napkins."

"I know it. How droll! And, why,
there is candy! Guy, Guy, there really
and truly is *candy !*"

"Didn't you ever see any before?"
asked Guy, much amused.

"Not for dinner; and there is a big,
bouncing watermelon. Well, I guess I

am very hungry," said Butterfly, subsiding at once, and becoming very quiet.

"So am I. I could eat—oh, there is Hal! I wish he was here. See, he has dressed for dinner, and so has his tutor. I like Mr. Niles, don't you? If it wasn't for Bertie, I would ask my father to let me have a tutor, and I wouldn't go to school, but I must now. There he is coming. I'll crowd a seat next to me. No, that isn't necessary. Mr. Niles has paid the waiter, and is pointing here. Now you will see him walk in without any trouble."

"Paid!" said Butterfly. "What for?"

"Why, for a seat, don't you know?"

Butterfly was too much taken up looking for Hal to answer. She could hardly understand the way in which the waiter

moved the elderly lady and gentleman
from the place they had already occupied,
putting Hal and his tutor into it, as if it
was theirs by right.

"I am glad you have come," said Guy,
cordially. "Sit here all the time, won't
you?"

"Yes," said Hal. "Let us be firm
friends and true to the journey's end."

Then the boys held each other's hand a
few minutes, sealing the bond, much to
Aunt Bessie's amusement.

Such a clatter of dishes tumbling over
the table, calling to waiters, taking everything within hand-reach, and pushing with
spoons to take hold of those dishes that
were not! Such confusion and bad
manners Butterfly had never seen before.
I think she must have gone away from the

table as hungry as she came, for she was too busy looking to eat.

Margaret was not at this table; and as Butterfly with her party went back to the deck after the dinner was over, she saw her, with her grandmother and a few others, seated at a small table in a part of the saloon which was curtained off from the rest.

"That is because her father is Lord John, I suppose," whispered Guy, following the direction of Butterfly's eyes. "She is too good to eat with common people."

"I think she is," said Butterfly, simply. "I love her dearly, don't you?"

"It will do for girls to love every pretty face they see, dearly," said Guy, with what he thought was a manly shrug of his shoulders; "but it won't for us boys."

"Why not?"

"Oh, because it won't," answered he, conclusively. "That isn't our style."

Now, Butterfly had only a very vague idea of what Guy meant, and it was just as well, for probably that was all Guy had himself; and as for its making any impression upon her that Margaret was or felt herself to be better than others because she was a lord's daughter, Butterfly was so truly a little American girl that I doubt whether she even knew what a lord was.

Dinner was at three o'clock, so that after dinner the afternoon seemed to the children well spent. The steamboat was now at a distance from land, speeding away over the still blue waters as fast as

7

steam could carry her; and as there was
very little to see, they went together into
a quiet part of the deck, and began to
divert themselves by that never-ending
amusement, story-telling. Now and then
Hal would call out that there was a town
on the shore, and occasionally a boat, a
schooner, or a ship went drifting by them;
but their main occupation was with their
stories, until the day faded away and
night came on.

All through the day Butterfly had
singled out from among the passengers
two who interested her. One was a pale,
sickly-looking young lady, almost half of
whose face seemed to her to consist in
large brown eyes. And the other was a
German, resembling, she thought, her
German friend who had travelled with

her from Lake George and then given her the ticket to the concert. The German on board the boat to-day had no violin, but Butterfly felt sure she heard him humming tunes. The lady, too, would break out into a line of a song, and then, as if suddenly recollecting where she was, would check herself.

When it became too dark on deck to see anything but the sky thick with stars overhead, and a long, dark expanse which Butterfly knew must be water around them, the children went down into the brilliantly lighted saloon. The first thing Butterfly heard was music, and the first thing she saw was the German sitting at the piano playing, and the lady standing by his side singing.

"I knew," she said, turning and putting

her arm around Margaret's neck, "something beautiful was coming. I've been looking and looking at them all day, and here it is. Where is Aunt Bessie? O dear, where *is* Aunt Matilda? Guy, run for your mother; and Hal, you bring Mr. Niles; and Margaret, bring your grandmother; and then I guess we shall be all fixed. We are going to have a real, live, true concert, without any tickets either!"

Concerts were not so much of a rarity to the other children as they were to Butterfly, so Guy said:

"Oh, they will come if they fancy. If these people are going to play a very lively tune, I should like it, or if they sing a merry song; but if it is what they call classical music, such as those Germans usually play, I wouldn't give a fig for it."

It was what Guy rather contemptuously called " classical music," so, after listening a little while, till her friends had joined another party of children in a distant part of the saloon, Butterfly went to see what was going on. She found them engaged in a play that was entirely new to her, so she sat by watching, no one, not even Guy, taking any notice of her, until her eyelids began to grow very heavy.

If you had told her she was sleepy, I think she would have indignantly denied it, but she put her head down on a marble table, and—well, I don't know as she would like to have me tell tales, but I am quite sure she did not for the next half hour get any very intelligent idea of the game. The first thing of which she was conscious was Guy's pulling one of her curls, and saying:

"Hullo, Butterfly, come on now, we want you!"

"We want you! O yes—dear me—yes—Aunt Bessie—no. Who is it, Guy? What?"

"I do declare, I believe you have been asleep," said Guy.

"Asleep!" And Butterfly slid down on her feet in a minute, rubbing her eyes violently. "I haven't been asleep, Guy Harrington. You ought to be ashamed of yourself." And only half awake now, Butterfly felt the tears coming into her eyes, and began to wipe them away as fast as she could.

Guy looked at her in much amazement. During all the time they had been together, he had never seen her cry or heard a fretful word from her before.

"I'm sorry," he said, in a very puzzled way.

But the tears had washed all the sleep away, and almost before the words were out of her mouth Butterfly was herself again.

"I don't want you to be sorry," she said. "I guess I was cross; I am sometimes. Now I am ready. What is it?" And really refreshed by her nap, Butterfly entered into the amusements more heartily than any of the other children for the remainder of a long evening; and when, quite late for a little miss who was used to early hours, Aunt Bessie came to say it was time to go to bed, Butterfly could hardly believe it was, as she said, nearly ten o'clock,

The musicians were still playing, and though many of the groups in the saloon had disappeared, yet enough remained to make the scene gay and cheerful. I do not know but Butterfly would have been willing to sit up all night if she could, and if Aunt Bessie would have kept her company, but her aunt led the way directly to the state-room, having first taken up a bit of a tin candlestick, with the half of a tallow candle in it. This she lighted with a match when she went into the room; and the first thing Butterfly did was to sit down and laugh at the funny shadows it made on the wall. Just one bit of a tallow candle was a new light for her to use, and perhaps things looked a little droller than they really were.

"Now, Aunt Bessie," said Butterfly, after

a few minutes' silence, " we must have our
Saturday's review, you know, and this is
the nicest place to have it in I ever saw.
I thought it was cunning before I went on
deck, and I think it is a great deal cun-
ninger now. I shall sleep as sound as two
tops, you see if I don't; and I want to tell
you all about Margaret's mother. O dear!
what if there should be a storm"—this
was the first time Butterfly had thought of
the possibility—" and it should make the
waters rise and the winds blow, and they
should lock us all down in a dark room,
dark as pitch? I wonder if there would
be any little girl to ask, 'Is Jesus here,
mother?' and then the mother would say
so pretty, 'Yes, darling.' "

" What is my Butterfly talking about?"
said Aunt Bessie, looking at her,

"Oh! I forgot. Margaret's grandmamma told us a beautiful story about being wrecked on this very lake; and how her little girl kept comforting everybody by telling them about Jesus, and how a man's voice out of the great darkness said 'Amen.' I should have been frightened almost to death; but she went first into the boat, not a bit frightened, because she thought Jesus was there too; and then the ship went down, and the water went over it, and Jesus was in the boat, so they all got safe to land."

"That was Margaret's mother and grandmother?" asked Aunt Bessie, much interested.

"Yes, it is true, too," Butterfly said. "I like the true stories best. I had better undress me, and say my prayers as fast as

I can, Aunt Bessie, so to be all ready if a storm should come ; but I guess there won't, don't you ?"

" There is every prospect of a beautiful night," said Aunt Bessie. " But, darling, it is always best to commit ourselves to the care of a loving Saviour's arms, to feel we can go to sleep in them quietly, as you do so often in mine."

" Yes, ma'am," said Butterfly, reverently. " I think Margaret did, that is, the one that was little Margaret then. Now I'll say my prayers.

Whether Butterfly, touched by the simple story she had heard, showing her how beautiful the trust of even a little child can be in her Saviour, determined to offer an unusually long and fervent prayer that night, or whether she was so tired that

closing her eyes even for a moment over-
came her, I cannot tell; but I know that,
after waiting in stillness for some time,
Aunt Bessie looked around to the spot
where Butterfly knelt, and found her, with
her head in her little folded hands, sound
asleep.

VI.

UNT BESSIE was wakened the next morning by a merry laugh close at her ears, and, opening her eyes, she saw Butterfly, with her head hanging over the side of her berth, holding on by both hands.

"Oh, Aunt Bessie!" she said, the moment she saw her aunt's eyes were opened, " this is so funny. Dear me, I thought I was at Frostland, up in my room, and when I opened my eyes, here I was—such a bit of a place. And I've slept splendidly; and I want to hop up and dress me just as quick as I can. I dare say we've been and gone through those thousand isles while I was sound asleep; and " —one little pink foot coming over the side of the berth—" I "—swinging down half over—" want "—landing on the floor, forgetting she was at sea, and getting a roll to the opposite side of the state-room.

"To pick yourself up," said Aunt Bessie, laughing, and finishing the sentence.

"Yes, ma'am." And Butterfly scrambled on to her feet in a moment, coming with a lurch up against Aunt Bessie's berth, who, putting out both arms, held her fast.

"I think," said Aunt Bessie, " the wind must have freshened during the night, or we were just stopping."

At this moment the steam-whistle sounded, and she added:

"Yes, that was it: we were stopping."

"This Montreal?" asked Butterfly. "Past the thousand isles and all?"

"Probably not in the St. Lawrence yet, so hurry and dress, for Guy is an early bird, you know. I dare say you will open the door and find him in the saloon waiting for you."

Butterfly hardly needed this prompting, but it would have been difficult to say which moved the fastest, her fingers or her tongue, as she dressed.

"I should like, Aunt Bessie, to read a chapter in the Bible, this morning, out of my common course. I don't think mother would mind it just for this once, if I read about Jesus on the water instead of reading with them at home. Do you?"

"No, I think she would be pleased to have her little girl associate Jesus with the water. Let us look at the chapter together. Do you know where to find it?"

"I know the story all about Peter and the other disciples out in the storm, but I guess we had better read it. I like Bible stories best. I think it is in Luke somewhere, and if you will wait I can find it."

Aunt Bessie waited very willingly until the toilet was finished, then Butterfly seated herself with her Bible, and after a few minutes said:

"It is in Luke, eighth chapter, beginning with the twenty-second verse. May I read it aloud?"

"Do, I should like to hear it."

Then Butterfly read every word of the chapter slowly and carefully, as if she was enjoying it. Then she asked Jesus to be with her all day, not as if she thought he was a distant Saviour who lived away off in the heavens and neither saw nor cared for a little earthly child, but as if he was a near friend who would love and bless her.

"Now good-by, Aunt Bessie," she said, when at last, ready to go on deck, she came to kiss her. "I am as happy as a queen, and I am going to have the most splendiferous time that ever any child did have ; and I am going to tell Aunt Matilda good-morning, so you won't see me until we get to Montreal."

"What, not at breakfast-time?" asked Aunt Bessie, kissing her.

"Oh, breakfast-time! Are we to have our breakfast here too? I'll go and hunt up Guy this very minute, and tell him all about it."

No sooner had Butterfly opened her state-room door into the saloon than Aunt Bessie heard noise enough to assure her a party of children were awaiting Butterfly's appearance. She could distinguish the voices of Guy and Hal, and, she thought, of Margaret also.

"She will be happy now," said kind Aunt Bessie, smiling all alone there to herself to think how contented Butterfly would be.

"Well, if you are not a sleepy-head!" was Guy's first salutation. "Here Hal and I have been waiting twenty minutes by my watch for you. I tell you what,

you would have got a drumming up if I
hadn't been afraid of waking your aunt.
Margaret has been on deck three times
and has come back for you, so, old lady,
hurry!"

"Hurry!" Butterfly felt as if she could
fly just as easily as not, if her wings were
only of the kind that would bear her up.
She skipped along over the saloon and up
the stairs so lightly that I think her
motion was about as much like flying as
it could be, considering she was only a
little girl.

When they came on deck, it was so
early in the morning that very few passengers were around. The ship looked
long, clean, and empty. Butterfly could
hardly believe it was the same place that
had been crowded and noisy on the

night before. Every one they met said
"Good-morning" to them so cheerfully
that Butterfly could hardly contain her
delight. I am not sure, if she could have
done as she liked, but she would have
stopped and kissed those who spoke.

"Now," said Guy, taking, as usual, the
control of the party, "we are to come,
before a great while, into the river St
Lawrence."

"We are in it now," said Hal. "Mr.
Niles, as soon as he found I had been
dubbed Guide-book, took and crammed
me."

"Crammed you!" said Butterfly, growing suddenly very sober. "What is
that?"

"Oh, nothing that hurt. When he tries
to make me learn a good deal in double

quick, he calls it 'cramming me.' That is an English phrase, I suppose," looking at Guy.

"We use it a little, but not much," said Guy. "I suppose it comes from cram-ming "——

"With anything; and I would not recommend it as a good word for you to copy," said a pleasant voice from behind them. Turning round, they saw Mr. Niles.

"I think it means just the same, and sounds better, if we should say, 'Gave me all the information he could.' Now, Hal, let us hear to how much purpose the 'cramming' was. But first look there, children. There is the sun rising. See how differently it looks, coming up out of

its water-bed, from what it does rising behind hills or from a long, level plain."

The children turned in the direction
from which the sun was rising.

There it was with its broad yellow disk
just above the water's edge, as if it had
been in for a morning bathe, and was
dressed in its golden robes to meet the
day. All around the little waves "danced
up to kiss it," Butterfly said. And I
think perhaps they did, for they looked
very happy and bright in its rays.

"There is not a finer sight in this world
than a sunrise at sea, my father says,"
said Guy, "and this is the first I ever saw.
I wish my mother was here." Guy
always thought of his mother at such
times.

"It is worth any one seeing. Let us

sit down where we can watch it. We have a great deal to see to-day that we ought never to forget, and this is a fine beginning.

The children all gathered around Mr. Niles, and after a few minutes he said :

"Hal, there is Kingston, and there also is Wolf's Island."

"Kingston and Wolf's Island," said Guy, getting up and looking about him as if he had been watching for them all the time. "What is it called Wolf's Island for? Now for a story, Butterfly. It is one of, the Thousand Islands."

"Yes, and I think it is one of the largest. By the way, to begin with, there are more than the famous one thousand islands. There are eighteen hundred, of all sizes and shapes, from a few **yards** square to miles in length; and if you like stories, there is one, or one can be easily made, for every island of the eighteen hundred.

"O Butterfly!" said Guy, his whole face lighting up with delight, "won't it be jolly?"

Butterfly gave her head such a number of assenting shakes that her curls flew about as if a lively kitten was in among them playing. Then she said, "Splendid!" and then Hal and Margaret smiled, as. if they thought her use of the word was a very funny one.

"Well now, about Wolf's Island?" asked Guy.

"First, of Kingston," said Mr. Niles, looking at Hal.

"Yes, information first, amusement

afterwards. That is what my father says, sir," said Guy. " I like them both, and pretty well mixed."

" Kingston used to have a queer Indian name," Hal began. " I shouldn't like to try to pronounce it, so I will spell it for you, and you can call it anything you please. C-a-t-a-r-a-c-q-u-i. The French once owned it — that was when they owned Canada—and here they built a fort. Then the French and the Indians used to try and see who should own it. Sometimes one did, and sometimes the other; but by-and-by it fell into the hands of the English, and then, of course, there was no more question to whom it should belong." said Hal, drawing himself up, proudly.

Guy saw the feeling in a moment, so he

said quietly : " There wouldn't have been if America had taken it. You changed the name from Cata—— something to Kingston, in honor of your king, I suppose."

" Yes, and now it is one of the most important military posts in Canada."

" I understand. Quite a little city, is it not ?" And Guy, accustomed to crowded New York, looked upon it, as he said, as a very little place. "Now for the wolf story."

A look passed between Hal and his tutor, which Guy noticed. And then Hal began :

" That island, you see, is long and narrow. Well, about half way down you will find an odd house. It was built a hundred years ago, I think, and there once lived a Frenchman, whose

name was Edward La True. O no!"
said Hal, stopping a moment. "He lived
on another island. Here lived an Indian
family, belonging to the tribe that once
owned Cataracqui. There was the father,
who was a famous hunter."

"That's good," said Guy, approvingly.

"And the mother, who did all the work,
and supported the family besides by sell-
ing trinkets."

"That's bad," said Guy.

"And four pickaninnies."

"What are those?" asked Margaret.

"Children, of course. Pickaninny is
an Indian name for a child. The family
lived all alone on this island. And it was
so full of game of all kinds that, if the father had been willing to take a little
pains, he might have gotten together

more than they could live on, without the least trouble."

"Lazy fellow," said Guy.

"But he used to trade off what game he took for liquor, and come home beastly drunk, to abuse his wife and children."

"He should have joined a temperance society," said Guy.

"He didn't. He looked like a prince when he was sober; and he felt like one when he was not, that is, he ordered his wife and children about, as if he was a world better than they were, and sometimes they had a very hard time of it. One day he came home, and told his wife to send Big Eye—that was his eldest boy—off to a distant part of the island, for some game he had killed and was too lazy to bring home."

" Did he go?" asked Guy.

" O yes! He went, and they waited and waited and waited; but he didn't come back. His father was hungry and very cross, so when the sun began to go down, and there was no Big Eye, he looked scowling enough, took up his bow and arrow, and without saying a word went out.

He was gone all night; and when the day broke, there was the mother sitting on the ground, looking out in one direction, just as she had been doing since dark; for soon after Big Eye's father went away, she heard a sound which she knew very well was the cry of a wolf near by."

" What did it sound like?" asked Butterfly, coming closer to Hal.

" Oh, I don't know. I have never heard one. It was between a lion's roar, a tiger's growl, and a hyena's call, I imagine. Something dreadful, at any rate. Now, she knew the wolves on the island were very large and fierce, and she had heard her husband say often that he thought a child would be in great danger from them, they were so wily and determined. She was afraid Big Eye had been eaten by one, and his father was in danger from the same. As soon as she could see, she took down a bow and a quiver of arrows, and off she started to help them. The island, as you see, is a very large one, but she knew where the trails were which they would be likely to follow, so away she went, as brave as a "——

"There is some one wants you, Guy."

"It's my mother," said Guy, and went at once, in the midst of the wolf story.

VII.

THE ST. LAWRENCE.

UY'S mother said to him, as soon as they were in another part of the boat :

"Guy, I am sorry to call you away; but there is an old lady here in trouble about some lost luggage. I thought you might help her by looking it up."

"Hal was telling me a first-rate wolf story," said Guy, annoyed when he heard for what he was wanted. "Wouldn't it do as well by-and-by?"

"I think not. She has mislaid a basket, and I presume, from her anxiety, it

contains something she would be sorry to
lose."

"O dear! mother, am I the only boy
in the world, that I must do every old woman's running that comes in your way?"

"You are my only boy," said Guy's
mother, looking kindly in his troubled
face; "but if it is so very irksome to you
to do a kind thing, I will call one of the
waiters."

"It isn't that, mother," said Guy, in-
stantly ashamed of his pettishness, "and
I am sorry for being so impolite as to call
her an old woman. I will go at once;
but I do wish I had heard the end of
Hal's story. The mother was just off
with her bow and arrow for the wolf, that
was growling and eating up her husband
and son at the same time. The old fellow

was selfish, and I don't much care for him; but I dare say Big Eye was a nice fellow."

By this time they had reached the person who had lost her basket, and Guy was quite himself again; but the old lady! Guy felt as if she *was* an old woman, without any of the lady, as he looked at her.

She was large, with great, coarse features, a mottled complexion, and eyes so small and nearly shut up by her fat cheeks that all Guy could think of, when he first saw her, was of a muskmelon, with eyes set in its ridges.

As soon as she saw him, she said, and her voice was loud and disagreeable:

"That is the little chap, is it? Well, he isn't much more than knee high to a toad;

but if he has got eyes in his head, perhaps he'll do."

Guy felt the color mount over his face, and knew the hot temper that was flashing through his veins had shown itself there.

" 'Knee high to a toad, and got eyes in his head!' A pretty way to receive a favor!" he said to himself.

If Guy's mother had not just then laid her hand gently on his shoulder, I think he would have turned, and without another word gone back to his story; but gentle as the touch was, it quieted him at once, and he said, with a self-control for which his mother gave him great credit:

"I am not very old; but if there is anything I can do for you, I shall be happy to do so."

The woman looked at him sharply from out of those little eyes, then said, curtly:

" You'll do. You'll find it, if it is on board, I'll venture you. It's a hand-basket, made by the Indians — blue, and red, and green, one of their brightest ones. I had it made on purpose to know it from other people's. So the first thing I've done, you see, is to lose it. It holds nearly a half-bushel, and is pretty heavy. It has a large card on it, with my name, Mrs. John Ragort. I'll be much obliged to you if you can restore it to me."

"I will do my best," said Guy, "but there is a great deal of luggage on board; and I saw quite a row of these large Indian baskets. They are so convenient, almost everybody who goes to Niagara purchases one."

"How like his father!" said his mother, watching him with much pride as he went away in the direction of the luggage-room.

To hunt up lost luggage on board a crowded steamboat is not a very pleasant task. Guy had travelled enough to know he must encounter all kinds of obstacles thrown in his way by every official who had any care on board; and so he did, but at last he was successful, and came back, bringing the very gay basket to its owner.

"I told you you would find it," she said, by way of thanks. And Guy thought she seemed quite as pleased with the idea of her own prophecy having proved correct as by the return of the basket. At any rate, these were all the thanks he received; but he heard her say, as he

walked quickly back to his young friends,
"That little fellow will make a man, if he
ever lives to grow up. He's a mighty
sight like my Josiah; and Josiah owns one
of the biggest farms in Illinois to-day,
and has about the largest drove of hogs
in the State."

Guy's mother with Aunt Matilda and
Aunt Bessie were now on deck, walking
up and down, and looking out the different
points of interest from a long panoramic
guide-book which they had just purchased. As Guy passed them, his mother
said:

"I see you have been successful.
Thank you, my son!" And then Guy felt
more than repaid.

"She says, mother," he said, pointing
carefully back in the direction where the

old lady was, " I'll do.   I am like Josiah,
who has the largest drove of hogs in the
State of Illinois."   Then with a merry
laugh he joined his young companions.

"Now, let me hear about Big Eye
and the wolf," he said. "I hope you
waited for me."

" No, I did not," answered Hal, "because the story was all gammon.   There
wasn't a word of truth in it, and Mr. Niles
says it was very poor at that."

"You are a most reliable guide," said
Guy, laughing.

" But there was an end," said Butterfly,
" if it wasn't true.   The mother shot the
wolf; and then she found Big Eye hidden under a big rock—Hal pointed it out
to us as we sailed by — so frightened
that he hadn't dared to move, hardly to

breathe, lest the wolf should find out he was there and eat him. He was so glad when he heard the arrow go whir, whir, through the air, and saw the wolf fall dead."

"And the father," said Guy, "the selfish old fellow?"

"Oh, he was saved too. The wolf had just got one claw on him, and his mouth all open to take a big bite, when whir, whir, went the arrow, and away he went after it."

"That is all, is it? Well, that will do for a make-up wolf story. Now, let us hear some true ones about these islands. Oh! look there. Isn't that a beauty? See, it is wooded down to the water's edge, and there in the midst is an old pine-tree that must have been struck by lightning some day. It looks like the mast of a man-of-war."

"That is prettiest, though," said Butterfly, pointing to a bit of greensward that looked like an emerald dropped on the smooth top of the water. "Oh! I do wish the boat would stop just long enough for me to build a baby-house on it."

"You'll wish so of every pretty island all the way to Montreal," said Guy.

"Yes," answered Butterfly.

"And if the boat were to stop; only think what a bother there would be on board," said Hal.

"Why?" asked Butterfly.

"Why, do you suppose all these men and women care anything about baby-houses?"

"No," said Butterfly, truthfully.

"I should like to explore fifty out of the eighteen hundred," said Guy. "Hunt up the animals and the brooks."

"It's famous hunting and fishing almost anywhere you choose to land," said Hal. "Mr. Niles and I always camp out through the month of September, and we never have been twice on the same island. Look there — see that tower. That is Martfield Tower."

"What is it built for?" asked Margaret.

"For some good purpose, you may be sure, for there is the English flag flying from it. It's on Cedar Island—named for its beautiful cedars—and there is a village a little distance from it."

"And two villages on the American side to one on the English," said Guy, good-naturedly, pointing to some clusters of houses near the shore.

"Is this Lake Ontario now?" asked Butterfly, suddenly. "I thought the Thousand Isles were in the river St. Lawrence. Dear me, how I have it all mixed up! We were in Niagara river, and then Hal said, 'Here we go into Lake Ontario!' And then I have not heard anybody say, 'Here we go into the river St. Lawrence.'"

"One would think Butterfly expected some great change," said Guy. "What do you expect?"

Butterfly hung her head—she would not like to have confessed just what she did, and she couldn't say anything unless it was the truth. Guy was afraid he had hurt her feelings, so he drew his stool a little closer to hers, and whispered:

"I didn't mean to."

"You didn't," answered Butterfly, understanding him at once; and then she caught sight of a lighthouse just before her, and forgot everything else as she eagerly pointed it out.

"That?" said Hal, looking in the direction of her finger. "Why, that is only a lighthouse. You see, there are so many rocks and so many islands."

"Or the other way," interrupted Guy.

"Or the other way," repeated Hal, good-naturedly. "There must be some method of warning mariners where the rocks or the islands, with their rock-bound, shelving shores, are; and the safest, indeed the only way of warning is by these lighthouses."

"I've seen numbers of them," said

Margaret, "but the finest one was the Eddystone Lighthouse. We passed that once in a storm."

" Where is it?" said Guy.

" Oh, it is over by the English shore, on the south coast somewhere. This is so small in comparison," pointing to the one they were now passing.

" This is a little house," said Butterfly, looking at it curiously, " just like a tower, with a belfry and all windows."

"That is just it," answered Hal; "and where it is all windows, there they put a big light, which keeps turning about all night, and when the sailors see it turn they say, ' Here is such and such a lighthouse ; and close by it, on the right hand, is a reef of rocks, and north of it lie the five islands, clustered so close together that, if we get in among them, there will be no getting out with our bones whole.' "

"Bones! Why, I thought you were talking about a ship," said Butterfly.

" And they speak of the ribs of a ship, don't they, I should just like to know?" said Guy.

Hal smiled, but went on with his explanation. " The government hires people to live in these lighthouses, and take care of the lamp, to see that it never goes out after dark."

"I should love to live there dearly," said Butterfly, going to the side of the boat nearest the lighthouse and looking out. " Do look! There is the nicest little garden, with some real, live flowers, and a dog; and—O dear! Guy, there is truly a little girl, too."

" Let us wave our handkerchiefs at her," said gallant Guy. And away went the four handkerchiefs, fluttering towards the little girl, who, never supposing they were intended for a salute, stood, with her feet wide apart, staring at them.

The sail among the Thousand Islands, though at first it seems as if nothing could ever be more delightful, becomes wearisome. Passing through them for once, and once only, it is impossible to distinguish or to remember one from another. They are of all shapes and sizes, as I have said ; but, after all, there must be something besides wood, rock, and land to imprint them on a child's memory so that they shall not be forgotten. One of these groups seemed to Hal particularly worthy of notice for this story, to which, while he was telling it, the children listened with much interest:

"Once while there was some political trouble in Canada, one of the prime movers of the rebellion was captured, escaped, and took refuge among this group of islands. There was a large price set upon his head, so a great many persons were out in pursuit of him, anxious to obtain the money.

"Well, they got word that he was hidden among these islands, and they determined they would capture him; the poor man, when he found his retreat was known, hadn't the least idea in the world that he could escape. He gave himself up for lost, and was sitting one day hidden

under that very rock there "—pointing to a
rock—"when he saw a boat pushed to shore
near by and a young girl jump out. At
first he thought he must hide, but, looking
a little more closely, he saw it was his
young daughter, Lucy. He knew, of
course, that she had come to find him, so
he stole out under cover of the bushes,
until he told her by signs to come to
him.

"Then she came to him, hiding as he
had under bushes and rocks, so not to be
seen if she had been previously tracked;
and when she reached him "——

"She kissed him," said Guy, looking
provokingly at Butterfly.

"Yes," said Butterfly, innocently.

"I doubt whether she had time," continued Hal. "She came to warn him that

his retreat had been found out, and to carry him away to another place."

"She! How could a young girl?" asked Guy, in much surprise and with some contempt.

"That was the very beauty of the thing. She was such a young girl that no one suspected her of so much daring. She had always been a pet of her father's, and he was very fond of bringing her among these islands for pleasure, so she had learned every channel and was a bold oarswoman.

"Now, for weeks she came at night, and, when no eye could see her, conveyed her father from one place to another, baffling every pursuer, until, tired of hunting for him in vain, the authorities gave him up, and he escaped."

10

"O dear! how I wish I could row!" said Butterfly, looking longingly towards the islands.

"I'll teach you when you come to see us," said Guy. "My father owns a boat on the Hudson, and I go out just when I please."

"Do you row?" asked Hal.

"Of course I do."

"Then we will have a harbor row when we get to Montreal," said Hal, gayly. "There is Ogdensburg."

VIII.

THE RAPIDS.

T Ogdensburg our travellers changed the large and commodious steamboat in which they had sailed through Lake Ontario for a smaller boat, made expressly for the St. Lawrence passage. There was so much bustle during the change, that Hal could not gather his party together to tell them about this town until they were once more on their way. Even then he had to call several times before he could draw them towards the point, from which they could catch a last view of the town as they rapidly left it.

"I want to tell you," he said then, as they huddled together in the stern of the boat, with the pretty town rising behind them, "that 'way back, in 1748 "——

"More than a hundred years ago," said Guy.

"Yes, more than a hundred years ago, Abbé François Piquet came over here as a missionary to the Iroquois Indians. You see, here is where the Oswegatchie river empties into the St. Lawrence; and just in such places the Indians love to congregate. He built a fort, and did a great deal in the way of teaching the Indians; but they got to fighting, and that ended the whole—that was for the time. The white people had found it would be a beautiful place to live in, and came back to it in due time."

"Then, what makes it chiefly interesting is, because it was begun as a missionary settlement," said Guy.

"Yes, and that is worth knowing."

"Of course it is. My father says no knowledge is ever lost. Butterfly, where are you flying to now?"

Butterfly, without stopping to answer, ran quickly across the boat, the children following.

To Guy's surprise she ran directly up to an old lady, who, though he had not seen it, had been beckoning to her.

"One of you is enough," said the woman, gruffly, as she saw them all coming. "It's the little girl with blue eyes that I'm wanting."

The others, not used to being thus addressed, turned at once; but the woman only winked at Butterfly, and seemed very much amused.

"Josiah would call that 'cutting and running,'" she said. Butterfly looked at her as if she did not quite understand what she meant, nor do I think she did. "Look here!" she then said, pulling her dress from over a large basket, which Butterfly knew at once to be the one Guy had found. "That is for him," pointing to Guy, "and I want you to carry it to him, formal like, with my compliments and my kindest thanks. Peep in!"

So saying, she lifted the lid of the basket, and Butterfly's eyes grew very large and bright as she saw its contents. Peaches and pears, plums and grapes, were arranged in the prettiest manner. Every peach turned with its red side uppermost,

and the grapes, purple and luscious-look-
ing, trailing like festoons around the pears.

"Those are for Guy!" Butterfly exclaimed.

"Yes, with my compliments and kind
thanks," repeated the woman, her coarse,
homely face lighting up with a pleasant
look.

"O dear me, how splendid!" Butterfly exclaimed. "I shouldn't wonder if he
was as happy as a king. Let me take
them, quick! Thank you very much indeed—that is, Guy will thank you, and I
do too—for—for"—said Butterfly, stopping suddenly—"for letting me carry them
to him."

"I'm thinking you'll get a bite, if my
eyes don't deceive me," said the woman,
looking very wise; but Butterfly did not

know what she meant. She only knew
that there was Guy, and here was the
basket, and the distance between them
was to be passed over in the least possi-
ble time.

The basket was as heavy as she could
well carry; and no sooner did Guy—who,
after his repulse, had walked a little
angrily away to another side of the boat
—catch a sight of her with so much more
than she could easily lift, than he ran to
meet and help her.

"They are all for you—the lady says
so"—placing the handle of the basket as
quickly as she could in Guy's hands—
"and there are lots of them. O my! so
splendid! Dear me! I never, *never*, in all
my life saw the like, and you are to have
her love, and her—no"—said Butterfly,

stopping abruptly — " her compliments,
and her love, and her regards—no, not
that either—her—her "——

"Never mind her what," said Guy,
peeping in ; " we've got her fruit, and that
is enough. See here, Hal, if a fellow gets
this for being cross and snappish and
rude as a bear, when he is asked to do a
favor, what wouldn't happen if he did it
as he ought, like a gentleman ?"

" Just nothing,'" said Hal, looking over
Guy's shoulder into the basket. " Mr.
Niles says we ought never to do a good or
a kind action for the sake of a reward."

" It makes a chap feel so mean," broke
in Guy. " I tell you I was as savage as
a meat-axe ; and when she called me ' knee
high to a toad,' I could have ——" opening
his hand in a very expressive manner, " if

she had not been a woman, and an old one at that. I can't take them, Hal— there is no two ways about it. I cannot."

" What can you do, then ?"

" Ask my mother." And away went Guy to talk the matter over with his mother. Pretty soon he came back with a very happy face.

" All right," he said, merrily. " Now we have nothing to do but help ourselves, and here is a rich treat indeed."

Guy felt very proud as he acted the part of lord of the feast. His mother had sent him to thank the lady for her gift, and had whispered to him :

" Let it be a lesson to you for the future. Always be obliging when you can, and be so whole-heartedly ; that is the beauty of the thing—*whole-hearted.*"

Guy had determined he would, and it was this good resolution that made him so happy now.

Before the children had quite disposed of the fruit, Mr. Niles came and said: "The boat is approaching the first rapids."

Now these rapids are, perhaps, greater objects of interest in sailing through the St. Lawrence than even the Thousand Isles, particularly to the boys. So the remainder of the fruit was put away hastily, and they all hurried to that part of the boat from which the best view could be obtained.

They were now about to enter the first rapids, called "Galliope Rapids." These are a very good preparation for what is to come afterwards. The water runs swiftly, tumbling up as it passes over the rocks, like waves, and when the steamer enters, it seems to fly along, as if it were ready to plunge down into an abyss.

The children were prepared, however, for something far more wonderful. Guy and Hal looked a little contemptuously upon what they were expected to admire; but they had both been at sea, and these first rapids do not seem great to those who have been tossed upon the Atlantic.

"Is this all?" said Guy.

"All! It's splendid!" said Butterfly.

"I think it's very nice," said Margaret. "I like it better than the others, there is so much less danger."

"I don't see any fun at all here," said Hal; "but you wait. If you don't sing a

different song, Guy, before we reach Montreal, then I am mistaken, that's all."

"I hope you will be," said Guy, with an
incredulous shake of the head. "If you
can believe it, Hal, my friend of the
fruit-basket asked the captain if he was
sure his boat is safe—if there is no danger
of its going to the bottom—where the
worst rocks are—and "——

"Little boy," said a voice close to Guy,
"when you are as old as I am, and Josiah
isn't with you, you will learn that it isn't
the prettiest thing in the world to be
a-going head-first you don't know where."

"But we do know where," said Guy, as
quietly as if he had known all the time
she was there. "We are going to Montreal,
and the rapids only help us to go the
quicker."

"Don't trifle with solemn things," said the woman, looking at Guy very sternly; and then she turned, and away she went, Guy's face changing into a very merry smile.

There were several small rapids through which the boats passed before they reached those which are considered the most beautiful in the St. Lawrence. These are called the Long Sault, and are nine miles long, divided in the centre by an island.

As soon as the boat came within sight of this white water, there began to be a movement on deck. All the passengers came up from the cabin, and those who were in the stern of the boat pressed eagerly to the fore part. Mothers called their children to them, and families everywhere arranged themselves in groups, as if expecting some great event, and being determined to meet it together.

Aunt Bessie would like to have had Butterfly close by her side, but Aunt Matilda said, sensibly : "If you show any anxiety, you will give it to her. She is perfectly fearless now, and will enjoy everything much better if she continues so."

Therefore Butterfly was left with her young friends.

"Shut off the steam, I see," Mr. Niles said, coming towards them. "Now, children, come with me to the wheel-house. Batiste is just coming on board. Did you hear them whistling for him?"

"Oh, that was it," said Guy. "I knew it was a signal, but I could not tell for what."

"There he is. Look!"

The children turned to see an Indian running up the rope-ladder that was lowered for him. Then his canoe, a long, slim boat, made of birch, was drawn up after him. He waited to see it safe. Then with a bound he sprang down on deck, and with long strides, without turning his head to the right or the left, or taking the least notice of the passengers who crowded around, he went to the wheel-house and took command of the vessel and of all the lives on board her.

Batiste was a large man, dressed with all the oddities of Indian costume. He wore tight-fitting pants made from a wolf-skin, the hair inside, a shirt of striped woollen, a sash of red broadcloth, elegantly embroidered with beads and figures made of colored barks, ended by a heavy bead fringe. A long, gray woollen blanket, embroidered also in showy figures, was fastened at his neck, with a round pin, cut from a stone, with a mosaic of steel. Long black hair, coarse and straight, hung down to his waist; and fitting tight to his head was a round cap, heavily worked with beads, and surmounted by two eagle plumes.

He had piercing black eyes, high cheeks, a large mouth, and looked so grave and stern that Butterfly crept close to Margaret, putting her arm round her waist as she did so.

Guy saw the action and said: " You are afraid, Butterfly. He has too much else to do to swallow you up at a mouthful."

Butterfly could understand very little of

11

what the others seemed to know so much. She only knew that Batiste took hold of the wheel, and three other men with him, and a little distance from him were two more, pulling away at something Guy called a "tiller," and then the boat went plunging down into the water, and rocked and tumbled about in a way that seemed very funny to her, while the faces of the six men grew red; and when Batiste grunted out a word, as he did every now and then, their faces grew even redder, and they pulled away as if they meant the steamer should go just Batiste's way and no other.

"We go like shot," said Guy, after watching their progress silently for a few minutes. "But after all, the waves are nothing like those of the sea in a storm."

"You wait until you reach Montreal," still said Hal, and only smiled when Guy looked so incredulous.

There was very little time lost in passing through these rapids, and long before Butterfly was ready for still water they were out of them, and Batiste was down among the crew, smoking his pipe as quietly as if there had never been such a place as the Long Sault Rapids.

Very soon after leaving these rapids, Hal said:

"There, Guy. Here the St. Lawrence says good-by to America, and becomes wholly her Majesty's!"

"And her Majesty is welcome to her," said Guy, with a low bow. "We have plenty of better rivers left."

Hal laughed. "You Americans can brag," he said.

"Yes, we inherited the trait," answered Guy. And Mr. Niles, when he heard his answer, laughed heartily.

"Now," said Guy, "I don't want to hear nor see a thing until we come to the Cedars. This quiet water is very tame." And he sat down, covering his face with both of his hands.

"Then you don't want to hear the story."

"Of what?" said Guy, uncovering one eye and peeping out.

"Of the old Indian village of St. Regis and the bell."

"I do," said Guy.

"No, I won't tell you until after we pass the Cedar Rapids," said Hal, provokingly.

"I will, though," said Margaret.

"Do," said Butterfly.

"Begin, then. 'Once upon a time,'" said Guy.

"Do it up short," said Hal.

"Catholic priests told the Indians who lived here, and whom they had half converted, that they would never go to a happy world hereafter unless they built a church and bought a bell. So they sold furs and raised money enough to buy a bell in France. But on its way to this country it was captured and carried into Salem, Mass. Then some people in Deerfield wanted to buy it, and they did, and carried it home, and put it up in the belfry of their church. When the priest heard where it was, he told the Indians they must go and find it. So away they went, through woods where there wasn't any path, until they came to Deerfield, in the night. Everybody was asleep, and they went right into the town, killed evei so many, and took others prisoners, and then they fastened the bell to a long pole, and carried it home."

"Dear me, I should think that was a new way, to murder for a church-bell," said Guy.

"They did, though," repeated Margaret. "I have been at St. Regis many times, and seen this very bell."

"Just done in time. Here we go into the rapids. Now for the same things over, but you just please to wait for La Chine, said Hal, "and then we'll talk about the *rapids.*"

IX.

COMING INTO PORT.

A CHINE Rapids fully justified all Hal had said in their praise. Batiste took the helm again, and the four men were at the wheel and two more at the rudder. All the passengers crowded to the fore part of the boat as soon as the roar and tumble of the water gave warning that they were approaching this famous pass. Our party of travellers, with the addition of the new friends they had made since leaving Niagara, Hal and his tutor, Margaret and her grandmother, gathered together, and fortunately secured one of the best places for seeing and hearing. On went the good, stanch boat bravely into the water.

"Just as if she knew what was expected of her," Guy said, "and meant to meet the occasion nobly."

Now came pitching and tossing, a feeling as if the boat was going head first into a watery grave—a sharp grinding when its keel struck against a stone, a little screaming from the most timid of the passengers. Pale faces and anxious looks on many that were not frightened enough to express their fears, and a general wish that the rapids were passed in the hearts of all.

As the boat went into the midst of the roughest water, Butterfly, who was much attracted by Batiste and the men at the wheel, saw the captain spring suddenly from the wheel-house down upon deck, and run towards the tiller. Two other men followed him in an instant, and then came screams from every part of the deck, and she heard the words :

" We're lost! we're lost !"

Catching tight hold of Aunt Bessie's arms, she looked around her in much alarm. What had happened?

Passengers began to push back from the fore part of the boat. Some covered their faces, some wrung their hands, a few dropped on their knees, and Butterfly heard the old woman, whose basket Guy had found, praying.

To her question of what had happened, Aunt Bessie could make no answer. Indeed all that was known was that the captain and men had jumped from the wheel-house with pale, excited faces ; and Batiste, left with one man only, was apparently doing the work of all.

Guy went to his mother's side, putting his arm affectionately around her neck. This Butterfly could see while she watched Batiste.

The ten minutes that followed, the child will probably never forget. The roll of the vessel, the roar of the water, the terrified passengers, Batiste and the other men that managed the boat, and Aunt Bessie's calm, pale face, that said to her as distinctly as the little Margaret once had to her mother, " Whatever happens, Jesus is here."

It was not until the peril was safely

past, that Butterfly, or indeed any pas-
senger, knew what had taken place.

One of the large ropes upon which they
depended for the control of the boat had
suddenly parted, and had not the prudent
managers provided for just such an
emergency, the boat would have struck
upon the rocks and many lives must have
been lost.

"Were we really in danger?" Butterfly
heard Aunt Matilda ask the pleasant
captain, when, the rapids passed, they
were again in still water, and he was
mingling with the passengers.

"Yes, great danger, for a few minutes,"
he said; "but, thank God, it is over.
These rapids are always dangerous, and
I never pass through them without being

inclined to say as I do now, 'Thank God.'"

"I like that man," said Guy afterwards, when the children were discussing their escape together. "He is a brave, and a good man too. My father says the bravest men are always those who can see God's hand in everything, and love to thank him heartily, as Captain Black did."

Hal looked at Guy in some surprise. He was not accustomed to hearing a boy speak of anything religious in a common conversation; but Guy had been brought up to recognize religion as nothing separated from, only the most important part of, his every-day life.

The children were now so fully occupied in imagining what might have happened, that they did not notice very many of the objects of interest around them, until Hal said :

" Look there, can you show anything to equal that in the United States?" pointing, as he spoke, to the Victoria Bridge.

Instead of answering him, Guy asked, " How long is this bridge? who built it? and for what special purpose?"

" Well, sir," said Hal, seating himself deliberately, and assuming the look of an old guide, "I will begin with your last question. The bridge was built by the Grand Trunk Railway Company. They must, you see, get over this river in some way, so James Hodges, builder, said to Robert Stevenson, engineer, ' I'll make a bridge for you,' and Mr. Stevenson said, ' Go ahead,' and so he put down twenty-four piers, two hundred and forty-two or three hundred and twenty feet apart, over a distance of one mile and a quarter. Now, in winter, you know, Canada is a cold place, and the ice is something of a somebody. It comes pressing down against this bridge, not like a trifle, let me tell you, but with a weight of thousands of tons. So Mr. Hodges made all the abutments of this bridge able to resist a pressure of seventy thousand tons. If we have time, we will go to the top of that centre tube. I tell you what, Guy, you will see things there that will make you stare."

"That's what I can do easily," said Guy, opening his mouth and eyes very wide, at which the little folks, of course, laughed.

"How much did it cost?" asked he then.

"That is a regular Yankee question," said Hal. "It cost seven millions of dollars."

"A pretty sum! I hope it pays."

"I think it is not good property." And Hal looked as concerned while he was saying this as his father would have done. "We pass right under it, in order to enter the harbor at Montreal, and you will see then how high it is above the water."

"Do we have to lower our chimneys? I've been often in boats that were obliged to do so in order to sail clear of a bridge."

"You wait and see," was all Hal answered.

"I see it," said Butterfly, suddenly.

"See what?" asked Margaret.

"Montreal, I guess."

"Is that all? Why, I thought it was something great," said Guy.

"It is," said Hal.

Guy smiled. He had been into the harbor of New York so often, and seen that great city spread out before him, that Montreal did not seem to him so very large.

Hal saw the smile, and answered:

"Here is a river frontage of nearly three miles in length, not easily beaten anywhere. For more than a mile it has a stone wall; and see there," pointing to a glittering dome—"there is Bonsecours Market; and there by it, you see, Guy"——

"Yes," said Guy.

"That spire belongs to the Bonsecours church, one of the oldest churches in Montreal. Look at Mount Royal, away there in the background. Don't you see those beautiful villas?"

"I do," said Butterfly. "I see houses and belfries."

"Spires," interrupted Guy.

"Spires," went on Butterfly, good-naturedly; "and—and—lots of things."

"Yes, that is just it," said Guy. "Lots of things. Now here we go under the bridge. I tell you what, Hal, you were right. This is a pretty tall sort of a thing. Look up, Butterfly. See what a distance there is above us before we come anywhere near it. It is grand!"

This hearty admiration Hal enjoyed, and I think the face of the English boy, as he stood looking up to the bridge above him, was made very beautiful by the love of his country which spoke in

12

every look. Guy thought so too; for in-
voluntarily he held out his hand, and
said :

"Hal, let us be friends longer than for
the voyage."

Hal hesitated for one moment. He had
not been educated into that love for
Americans that would make him take one
warmly and at once into his affections;
but there was something about Guy
which he could not resist, and he put his
hand cordially in Guy's, his grasp warmer
and longer than any other boy's had ever
been before.

"There is Notre Dame," said Margaret,
while they still held hands under the
bridge.

"Notre Dame? Who is that?" asked
Butterfly.

The children all laughed.

" *Who* is a large Catholic church," said Guy.

" Oh, I thought it was a woman," said Butterfly.

" It is named after one," said Hal, " ' Our Lady,' meaning the Holy Virgin, Mary, the mother of Jesus. We will go together and visit it, Guy. Mr. Niles says he will extend my vacation as long as you remain in Montreal, and if you will take me for guide, nothing will make me happier."

When Margaret heard this, she ran to her grandmother, and soon came back with a very smiling face.

" My grandmother says," she said, " that I may go too. May I?"

Butterfly, by way of answer, hugged

and kissed her, and Margaret seemed to.
understand that she meant, "Yes, it
would make her very happy to have her
with them," as in truth it did.

"What is that tower?" asked Guy,
pointing to a tall tower.

"Royal Insurance Buildings," said Hal;
"and all that long line of handsome stone
buildings flanking the wharf are stores.
We shall land at Jacques Cartier's
Square. Here we are, headed for it now."

The boat steamed slowly through long
lines of shipping until it came to a large
wharf, before which several other steam-
boats were already landing passengers.
Each one had its place, and so, carefully
avoiding any collision, went to its own
landing, let off its steam, blew its whistle,
rang its bell, and stopped.

Butterfly's aunts did not offer to take the few cares they had given to Butterfly from her because there was so much that was new to see and hear here. She would very gladly have stood still and left everything to be done by some one else; but she very well knew this was neither kind nor polite, so, with many longing looks behind her, she went down into her stateroom, and began to gather together all her travelling articles. These had accumulated since she left home. Indeed, I am afraid, if her kind aunts had not kept a quiet watch over them, she would hardly have had them all when she reached her own home in Frostland.

When she came back on deck now, she looked like the little old woman we have all heard about who never could travel without taking with her "big box, little box, bandbox, and bundle." So Guy told her, but she didn't care, only so she kept them safe.

"O my! dear me! how funny!" Saying these words, Butterfly began to drop first a basket, then stooping to pick it up, she let her sunshade fall. Picking up that, down went her reticule; picking up that, over went her water-proof; until she was only a little girl, sitting amidst a heap of travelling gear, trying to get it together with her hands, while her eyes were fixed upon a man dressed in a green coat, with white knee-breeches, who was coming on deck.

"Did you ever see such a funny-looking man?" she said, pointing him out to the others.

"Oh, there is John!" said Margaret. "John! John!"

The man in the green coat turned at once, then lifted a tall black hat with a white band around it, and came towards her.

"Who *is* it?" asked Butterfly, more and more astonished.

"Our footman, John," said Margaret. But Butterfly never had heard of a footman before, and did not know what she meant.

The man looked as if he was very glad to see Margaret; but she did not seem to care for him, and Butterfly felt sorry, and wanted to go to him and say a kind word. But just then he caught sight of Margaret's grandmother, and, touching his hat once more, went towards her,

Butterfly following him curiously with her eyes.

Pretty soon he came back, bringing Margaret a message from her grandmother, and then she said " good-by " to the children, promising to come early the next day to see Butterfly, at her hotel.

" Look out, now! " said Hal, as Margaret with her grandmother went over the broad board which had been placed between the boat and the wharf for the accommodation of travellers. " Look out for Lord John Talbot's carriage ! It's the handsomest turnout we have in Montreal."

Guy ran with Hal to a side of the boat, so that they could see up the street upon which the carriage had started, and Butterfly followed them.

However handsome the " turnout " was, the only thing which made any impression upon her was the two men, coachman and footman, dressed in what seemed to her the most ridiculous of dresses.

" Why, Aunt Bessie," she said, as soon as they were moving away from the boat in the omnibus that ran to the St. Lawrence Hall, " O dear! he looked just like —just like a picture I saw once in Emma's old English story-book."

" And I dare say he had on the very same dress that the picture was made to represent. These old English families seldom change their liveries."

" Liveries. What are those?"

" The dresses of their servants."

" How very, very funny!"

" That is only one of many funny things

you will see here," said Aunt Bessie; at hearing which, Butterfly had to whisper her delight to Guy, who looked very manly on the occasion, and said:

"I have no doubt we shall hunt up something in Montreal worth coming for, and to begin with, here is the St. Lawrence Hall, the very best here, and our home for the next week, at least."

X.

LETTERS.

T the St. Lawrence Hall a large number of letters were waiting our travellers. They were handed to Guy as soon as their rooms were taken, and he saw with much pleasure two for Butterfly and three for himself. As none of them are very long, and they are all from the young friends who have been with us in our preceding books, it may interest my young readers to have me copy them here. The first Guy opened was from Bertie, and must have been put into the office on the same day that the party left Niagara. Here it is :

" DEAR GUY :

" I miss you, old fellow. I wish you were here. I am counting the weeks already until we go to school together. This morning I am going to have the big black horse and trot down to the Suspension Bridge ; and this afternoon I am going to have another row up and down the river. We—that is the ferryman and I—are going up as close to the falls as we can. I mean to get in under them, so we can have a real ducking. But don't you tell anybody, for she—you know who— wouldn't let me, and then we should have a row. She and I are getting to be real good friends—quite affectionate—that is, she is, and I grin and take it.

"Good-by, parson prim. No, Prig, it was, I believe. Days fly by, and our time will come at last. Till then, yours,

"BERTIE."

"Good for Bertie!" said Guy, folding the letter slowly. "I should like to be with him in the boat. Mother, I wish I was the eldest of ten boys. Then you would let me go, wouldn't you?"

"No sooner than I would now, Guy," said his mother, smiling. "I think such exploits foolhardy."

Guy would probably like to have discussed this point if he had not two more letters to read; but while he had been busy with Bertie's, Butterfly had opened one of hers, and was now ready to read it aloud to him.

"Oh, it's elegant!"

"I don't see it. It looks rather crooked," said Guy, peeping over her shoulder.

"I mean—there—I promised Aunt Bessie I would try and not say any of those words to-day, and I have already. Well, it's from Marie, and it tells me everything.

"'DEAR, DARLING BUTTERFLY:

"'I love you so dearly, that if I could see you this minute I would give you 'most a thousand kisses.'"

"I wish you could," said Butterfly, interrupting her reading.

"'I go up to the Clarendon, and look up to your window, but you are not there.'"

"Evident, if you are in Montreal," said Guy.

"'And then I feel as if I should like to cry, and sometimes I do almost. Then

I go up to see Charlie. To-day I carried him some presents ; but mamma says it is not ladylike to talk about a gift you have made, so I will not tell you what they were. He likes them, and we went to walk together—that is, I walked and helped drag him—and when we got to Empire Spring, we hunted up the place where you told your story about Alice and Charlie told me another all about—but I can't stop to tell you now, because my letter wants to go to you soon, and I see Felice putting on her sash. Then she will say, " Come, Marie," and I must go.

" 'I love you dearly and I miss you always. I wish you would come back, so does Charlie, so do we all. Please come. I love you. I kiss you.

" ' MARIE.' "

Butterfly kissed the letter over and over again, and then Guy, laughing, said:

"That will do. Now here is one to me from Charlie's own self. Listen:

"' DEAR GUY:

"' Mother says all I can write, when I write to Butterfly or you, is: "I love you, and I want to see you;" but it isn't. I can tell you how Nap has come, and Marie knows him so well that he barks every time he hears her voice when she comes to see me. I tell you what, sir, he is the most splendidest dog ever was. You should just see him, if you want to know.

"' Marie's father came with him, and he laid my hand on his cold nose, and he said: "Nap, this is your little master, now. Be very kind and good to him, will you?"

And Nap made a little bark just as if he was saying, "Yes," only it frightened me so I took off my hand in a hurry, I tell you.

" 'Father says, when he gets to know me well, I may have him harnessed into my wagon and take a ride. But Marie's father gave my wagon a shake, and said, " It wasn't just the thing for a sick boy to ride in." And then he turned it up-side down, and said something to my father that I couldn't hear ; but when he went away, father began to tinker on it, and then it rode so easy.

" 'O Guy! Guy! Guy! you don't know what a beauty Nap is! With a black and white face, and a black and white body, and a white tail, all bushy, with a little black end, and four white feet, and such

13

ears! I wish you would come and see him—you and Butterfly. Can't you? Please do, and see, too, your loving Charlie, who is happy as a king.' "

"Three cheers for Nap!" said Guy, waving the letter over his head. "He is a real Newfoundlander, mother, like our old Tiger. I should like to see him right well, and to drive out Charlie. I say, Butterfly, let you and me go back to Saratoga."

"Well, yes, I should love to dearly. Oh, here, I have a letter from Charlie too."

"So you have. Read it, do."

" ' DEAR BUTTERFLY :

" ' Mother says I do nothing but want her to write to you or Guy all the time ; but Nap is so nice ; and what do you think ?

Yesterday a big, big thing came up to our door—I mean the express-man brought it—directed to "Master Charlie, from his friend Marie," and just as fast as we could, mother and I untied it—the strings, I mean, round the brown paper—and there it was. O dear! it is so nice I don't know how to begin to tell you about it. It's a carriage—not a wagon, you know, but a real, true carriage, all cushioned, and it's dark blue, and it's painted with gold—a little gold butterfly right on its side; and Marie says that is in honor of you, our own live Butterfly, and mother says 'In honor of' means that we love you and want to say so—and it's mine—my ownty-townty—can you believe it? I can't, but it really and truly is. "Charlie, from his friend Marie," you know.

" 'Then I've got a harness, a real harness, for Nap, just as if he was a horse; a saddle and a bridle, and hold backs, and O dear, such reins! Blue, with " Charlie " in great letters on them; and a whip—but I wouldn't touch it to Nap for the world. It's only to make it all up, like a real horse, you know. But I like a dog—I like Nap a great deal better.

" 'I am going to ride with it. Marie comes too, and father is going to show me how to drive. Good-by. I can't write any more, 'cause I guess Marie will be here soon, and I must go and get ready. Please tell Guy all about it.

" 'Your little happy, CHARLIE.'

" 'Mother says this is a pretty way to sign it—my letter, I mean.' "

Butterfly and Guy looked in each other's faces after this letter was read, with almost too much delight to speak. At length Guy said:

" Butterfly, that is what I call first-rate Only think, a dog, a wagon, a harness, and a whip! A complete outfit."

" And a gold butterfly," said Butterfly, dancing around, " because they love me so dearly. Oh, Aunt Bessie, I wish you would let us go right back and see it."

"Not this time," said Aunt Bessie. " We have Montreal before us yet, you know."

" We can't go, of course, Butterfly," said Guy, in a very old way, " but I mean to write to Charlie to have a photograph taken of the whole affair and sent to us."

"A photograph of a dog!" said Butterfly, opening her eyes very wide.

"Yes; and the carriage, with Charlie in it. I had one of Tiger. By the way, here is one in my pocket now." And opening his pocket-bock, Guy took out a picture of a large Newfoundland dog.

"What a beauty!" said Butterfly, holding it at arm's length.

"I rather think he is. You shall see him when you come. Father calls him, 'My inseparable,' when I am at home. If Nap beats him, he is a pretty fine fellow. That is the most I have to say. Now for my third letter. It is from Bennie: written in the cars, and mailed on the way. It's written in pencil. But see this, mother, don't you call it a fair hand for a boy like that?"

"I think it shows a great deal of character," said Guy's mother, looking at the note. "Are you going to read it aloud?"

"Of course I am.

"'DEAR GUY:

"'Hallo there! how are you? I've just been into the smoking-car, and there was a little fellow no bigger than you are.'——

"I am as large as you are, sir," said Guy, sitting up very straight.

"'Smoking his cigar. He said it was his third. What do you think of that? But I tell you what, his face was yellow, and all shrivelled up like an old man's I saw once who had had the palsy fifty years, I guess. I don't know, but it was some awful long time; and he—the boy—

couldn't stand straight. I thought he had been drinking, but it was only his cigars. He was weak in his pegs.

"'I am having a jolly ride. · We are in an express train, and go forty miles an hour.

"'I expect a long letter from you when I reach home. My mother bears the ride well; the girls enjoy it.

"'Write. With love to Butterfly,

"'Yours ever,   J—r!'"

"I should call your young friends very punctual correspondents," said Aunt Matilda, as the last letter was finished. "Now, how soon do you and Guy intend to answer these letters, Butterfly?"

"To-night," said Butterfly, promptly.

"No, in three days," said Guy. "Then

we shall have seen Montreal and have something to tell them about."

"But Charlie," said Butterfly, "I want to tell him how glad we are."

"True enough. You run to your room and I will to mine, and write him, then we will put our notes together in one envelope."

The children separated, and just before bedtime, tired as they were, they had written.two long letters to Charlie, which they put together, and Guy directed, with many flourishes of handwriting, to " Master Charles Wilson, Saratoga, N. Y." And then two happier children did not go to sleep in Montreal that night, I am sure, than Guy and Butterfly.

XI.

HAL ACTS AS CHAPERON.

HAT do you know about Montreal?" asked Guy the next morning, when Butterfly, looking rather pale and tired, made her appearance in their sitting-room.

"Nothing," said Butterfly, a little crisply. "I thought that was what we came here for."

"So we did; but don't you want to know who built it?"

"No," said Butterfly.

Guy looked at her in surprise. It was so seldom that Butterfly was fretful, he did not know what to make of it. "Then I shall not tell you."

"No, don't," said Butterfly, going to the window, and looking out.

Guy turned over the leaves of the book he held in his hand very slowly. He had been busy hunting up all the information he could find about the city, but what he had learned seemed very little reward if he could not impart it to Butterfly.

Conscious that she had been cross, Butterfly kept looking back at him from under her eyes, in a way she had when she did not wish to be observed, and then out of the window again; but she was silent for a long time, so was Guy. Suddenly she broke out with:

"O Guy! O dear! how funny! Hurry, hurry, Guy. Quick! O my!" And she jumped up and down, clapping her hands.

Guy was at her side in a minute, and saw—only a donkey-cart.

"What?" he asked, putting his head far out of the window, and looking up and down the street.

"That funny little cart and horse," said Butterfly, pointing down into the street.

"That isn't a horse at all. Why, Butterfly Courtland, did you never see a donkey before?" And there was something provoking to Butterfly in the laugh with which Guy followed this explanation.

"A donkey!" she said. "Why, I thought a donkey was"—— Then she stopped.

"Was what?" asked Guy. "Did you think a donkey was a horse?"

"No," said Butterfly, briefly.

" What then ?"

But Butterfly was now too fully occupied by another new thing to answer.

" Riding behind," she said, " driving over the top of the buggy !"

" Yes, that is a hansom." Guy wanted to ask again, " Did you never see one before?" but he remembered how he had annoyed Butterfly, so he said, kindly : " We have a few of them in New York ; but my father says they are very common in the streets of London. You see, the driver sitting behind does not obstruct the view of the person riding. It is just as good as if you were driving yourself, only you don't have the trouble of holding the reins. All you have to do is to look out and see what you can. We will have a

ride in one, Butterfly, before we are done
with Montreal, see if we do not!"

"That would be—O dear! I was going to, but I didn't, did I, Guy?"

"Did you what?"

"Say 'splendid.' There, I have now,
and I promised Aunt Bessie I would try
not to." And Butterfly looked discom-
fited.

"You didn't use it for an exclamation,
though. See, there is a line of donkey-
carts; and there comes another carriage—
a funny little single hack, drawn by one
horse; and—hush, Butterfly, I hear martial
music. I dare say some of her Majesty's
regiments are out. Come and see."

But though the music could be distinctly heard from every part of the house,
no soldiers were to be seen, and the chil-

dren, afraid of losing a sight, went for leave to go out; but Aunt Matilda was not willing to have Guy and Butterfly alone for the first time in the streets of a large city, and the elder members of the party preferred quiet rest in their room at home for this morning. So the children were just making up their minds to amuse themselves as they best could, with the sights to be seen from the windows of the ample parlors, when a servant coming in, with what Guy called a great flourish, said :

"Master Henry Tappan sends his card," presenting a small, neat card, " and if the young lady and gentleman are disengaged he would be happy to see them."

" Disengaged of course we are," said Guy, eagerly. " Where is he?"

"In the carriage at the hotel door," said the waiter.

"Tell him to come on," said Guy, very unceremoniously. Then, as soon as the waiter had gone, turning to Butterfly : "I dare say he will come with four men in knee-breeches and crimson plush, and turn out to be the Prince of Wales."

"Knee-breeches and crimson plush?" asked Butterfly, looking in Guy's face in a very puzzled way.

"Yes, that is the way the grand English people dress up their servants. Some people in New York imitate them— make popinjays of them."

"Popinjays!" repeated Butterfly, again, growing more and more puzzled.

"How funny you do repeat!" said Guy,

laughing. "Did you never hear of a pop-injay?"

"No, never; but it is no matter. There he is, and I don't see any knee-breeches or crimson plush."

Had Butterfly been any less glad to see Hal, there would have been a shade of disappointment in her face, as the boy came bounding into the parlor; but now the reception the two waiting children gave him was warm enough to satisfy any one.

"I've brought my father's carriage, and he places it, with his compliments, at your disposal for the day," said Hal, after a few minutes. "If you like, we will go out sight-seeing. I wanted to come in the dog-cart with the ponies, and father said I might, but it would be nicer to

14

have the carriage, if any of the ladies would like to go with us."

The invitation was so formally given, that the children were at first a little diffident about accepting it. They said they would ask leave. In a few minutes they came back ready to go, those in authority being very glad to have them sight-see under such favorable circumstances.

Butterfly was a little awed when she saw the handsome carriage and the two large horses; but perhaps less with them than with the coachman and footman, dressed, not in a harlequin style, but in plain black, with white gloves, and what Butterfly was sure must be a silver band around their hats.

"First to the cathedral," said Hal, as they were ready to start. "That makes a good beginning. O no—drive around the city, John. You know better than I do what strangers like to see."

"Very well, sir," said John, touching his hat.

The children now began to talk so busily that I doubt whether they would have remembered there was anything to see, if John had not looked back into the carriage, and said:

"Here is the Methodist church; it has one of the finest organs in the city."

"But it is never played, excepting on a Sunday," said Hal, "so that doesn't do us any good."

"Here is a Music Hall, holds a thousand people."

"Is that all?" said Guy, a little contemptuously. "We have"—— Then he stopped short. His mother had charged him before he came out not to compare a single thing he saw while with Hal with what he had seen at home. She told him it was neither courteous nor pleasant to do so.

" Molson's Bank," said John next.

" Oh! how handsome," said Butterfly, looking out with delight at its showy exterior.

" Caverhill Building."

" Stores, I see," said Guy.

"Drive through St. Paul's," said Hal.

The coachman touched his hat again, and turned into a street filled at this hour of the morning with all the bustle and crowd usually to be found in the principal business street of a large city.

"You should see Broadway," Guy was tempted to say, but he did not.

Butterfly laughed out every now and then as she saw something entirely new, but otherwise was too much occupied and amused to talk; indeed, the boys found it very difficult to get even an answer to their questions.

So many buildings did John point out in the course of the hour, while the great black horses were carrying them around the city, that Butterfly's mind began to be full of banks jostling churches, and churches pushing about universities, and universities crowding down nunneries, and nunneries turning into hospitals; so she was very glad when at last John stopped before a large stone building, and said :

"This is the cathedral."

"We must do that up, of course. I have been in here a hundred times, and if I live long shall be in as many more," said Hal; "but I am never tired of it."

The footman got down from the back of the carriage, and opened the door, holding his hand out to help Butterfly, but she jumped down like a little kitten, though she did not forget to say "Thank you."

When the massive door was opened, and the children stepped within the great church, Butterfly stopped short. She had been at church in Frostland all her life— a neat, simple little meeting-house, four like which could have been put easily within this.

But here was a vast building, so large that Butterfly, standing by one of the doors that opened upon the street, where their carriage had left them, could hardly see to the end of it, and instead of having only the four plain walls, a pulpit, and a little gallery for the choir, here were pillars upon pillars, galleries above galleries, pulpits everywhere.

"Pictures—real live pictures!" These were Butterfly's first words, catching hold of Guy's arm as she spoke.

"Yes, about as live as pictures ever are," said Hal, laughing. "Some of these are very fine. Look there, now. There is one of Jesus walking on the water: or no, I don't know what they are about, but they are all on sacred subjects. Mr. Niles says some are copies of the great masters. We will take them deliberately and do it up well."

"Do it up how?" asked Butterfly.

"That means, see it thoroughly," said Guy. "My father says that if a thing is worth seeing at all, it is worth seeing well."

"So does Mr. Niles, and he makes me, too. Now, come on. They are having mass at the other end of the cathedral. Did you ever go to mass?"

"No," said both children together.

"We will go up there, then, and see the rest afterwards," said Hal. "Never mind the people we pass. They are saying their prayers, but we shall not disturb them ; they look up, and go on just the same." So Hal led the way through the centre of the cathedral, the others follow-ing ; Butterfly meaning to walk on her toes and make no noise, but forgetting it every other step, in her interest, and in the end making quite as much as either of the boys.

Many of the kneeling people looked up, bowed, and smiled as the pretty child stole by them, but did not stop their prayers. To one old woman who bade her " good-morning " in French in between the words of her " Ave," Butterfly, supposing she was asking her a question, stopped, and said " Ma'am?" but the woman only smiled, and held out her hand

" She wants something," said Butterfly, whispering to Hal.

" Yes, they all do. You will have as many hands held out as you meet beggars in Montreal. The city is full of them."

" Beggars !" said Butterfly, turning

round and looking at the woman. "She is in church."

"That don't matter. They beg praying, and eating, and drinking, waking and sleeping, I believe. It's nothing but beg, beg, beg."

"Poor things!" said Butterfly.

"Lazy things," said Hal.

"Are they?" asked Butterfly.

"Yes; there is work enough everywhere to support everybody who is not sick, my father says," added Guy.

The children might have occupied themselves some time longer, discussing this somewhat difficult question of pauperism, if Butterfly had not at that moment caught sight of a priest swinging a censer.

"O dear, Guy, look! That man dressed so queer swings something; and do see

the people bow and courtesy. What
for?"

" That is part of the Catholic worship,"
said Hal. " Now watch them. We will
sit down here."

So the children seated themselves near
the part of the church where the services
were proceeding, and Butterfly, who had
never been within a church of this kind
before, was very much interested in all
that was done.

She, of course, was too young to understand what theologians might condemn.
She saw and heard what was passing, and
nothing more.

After the mass was ended the great
organ in another part of the church began
to play, and they hurried to the place
where it was; Butterfly pulling Guy's

sleeve to have him look at this, that, and the other thing, all the way.

I don't think the child carried home to her aunts, after a long morning spent in seeing, any very distinct account of what she saw; but she did remember, in this noted cathedral of Notre Dame, the pillars, the pictures, the little rooms set apart for the confessional, and, more than anything else, the size and solemn stillness, which seemed to her almost like separate things.

After seeing the interior of the church, Hal led the way into the tower of the cathedral; and Butterfly found herself climbing up, stair after stair, until, as she told Guy, she expected to come out among the stars; but she was only above the city of Montreal, where she could see not only the whole city, but much of the surrounding country.

"That is worth coming to see," said Guy, with admiration. "I thank you very much, Hal. When you come to New York, I shall be glad to show you our best, though for a view I do not think we have anything to equal this."

"Then we will end our sight-seeing to-day," said Hal, "or rather, this morning. After dinner, if you would like, I will come with my dog-cart, and drive you round the mountain. That is *the* drive of all others in Montreal. Back to the St. Lawrence Hall, now, John!"

So the black horses trotted in their stately way once more to the hotel, and the first person Butterfly saw as the carriage stopped was Aunt Bessie, looking out of the parlor window in search of her.

"I am glad to see you back," said Aunt Bessie, smiling and nodding.

XII.

GOOD-BY.

BEFORE the hotel dinner was through, Hal made his appearance in his dog-cart at the door. Butterfly had expected to see a carriage similar to the one Charlie had described as being a present to him from Marie's father; but to her surprise, here was only a small wagon, with two bay ponies harnessed into it.

"Oh, how beautiful!" she exclaimed, stopping on the steps and clapping her hands, much to the amusement of the lookers-on. "A real live dog-cart, drawn by ponies ; and "—catching sight for the first time of a little girl seated demurely on the back seat—" and Margaret too!"

One bound down the remainder of the steps, and Butterfly was clambering into the cart, with no other idea but the one of kissing Margaret as soon as she could.

The boys laughed, so did the man who stood at the head of the horses. Margaret heard them, but I question whether Butterfly did, or would have cared if she had.

A very merry-looking carriage-full they made. As they drove down the street many turned to admire the ponies and cart, the four happy children, and the two servants on horseback—one, Hal's man; and the other, the footman in the green coat and knee-breeches, whose looks had so much amused Butterfly.

The ride around the mountain was a fine one. Hal and Margaret knew everything there was to be seen. I should like to take my young readers there also, but our story is now as long as it should be, and so I must pass over this and many other objects of interest which the whole party visited, to a nunnery where children without any fathers or mothers are taken and brought up in the Catholic religion.

When Butterfly was told they were to go to a nunnery, Aunt Bessie explained to her that the nuns did not dress as other people do, but put themselves into straight black dresses with little tight sleeves, and a white neckerchief pinned over the waist of their dresses. Then they had

15

their hair cut off, and wore an odd cap with a big crown and a deep, white frill. In the picture Butterfly once saw of a nun she had her hands crossed over her breast, and a rosary round her neck, with a cross on the end of it. Butterfly thought she had a very sad face, and had dreaded going where she should see any living, until Aunt Bessie told her that the sisters who belonged to the nunnery they were now to visit had collected a great many little children together whom it would be pleasant to see.

Butterfly remembered the " Home for Destitute Children" in Burlington, and what a nice time she had in visiting that ; so she summoned all her courage, as she did when she was going to see the Indians, and went up the steps and into the long, still halls, without saying to any one how afraid she was.

They were taken directly to a schoolroom not unlike many others, where the boys and girls old enough to be taught were ranged, the boys on one side of the room, the girls on the other; and here were four of the nuns as teachers. One of them, as soon as she saw Butterfly, came to her and took hold of her hand. Butterfly trembled a little as she did so, but when she found it was warm and tender, " almost," she said to herself, " like Aunt Bessie's," all fear left her, and from that moment she began to enjoy herself.

" The children look so oddly," she whispered to Guy the first time she could come near enough to him to speak.

" The boys look as if they had walked

out of a primer," said Guy ; "blue homespun, I know."

"And the girls," said Butterfly, "look
like the picture of John Rogers' children,
when he was burned at the stake, you
remember."

"Yes," said Guy, "so they do. Nine
of them : long blue homespun again, short
waists, short hair, wooden shoes."

"Wooden shoes !" said Butterfly.

"I should think so. Look there, now!"

A little boy walked across the room as
Guy spoke, and his feet went tramp,
tramp, tramp, precisely as they would
have done had his shoes been, as Guy
said, "made of wood."

Some recitations were carried on while
the children were there, but in French,
and very drony even in that lively lan-

guage. Guy thought his teacher would
have said, " Wake up," and I rather think
he would.

After visiting a variety of other rooms,
kitchen, laundry, dining-rooms, where Guy
pointed out the tin pint cup and wooden
spoon with which each child large enough
to go to the table fed itself, they came to
the room where the babies were; and of
this Butterfly will never be tired talking,
for there were fifty babies here, all dressed
alike, and, as she said, over and over
again :

" All dressed just like play-servant
dolls, all but the caps."

These little creatures of every age,
from two days to as many years, wore
blue homespun dresses, loose, with just a
string around the throat to draw them up

close, long sleeves, snug to the tiny arms, and on their heads little white cotton caps without any frill, just fitted tight to the head, and tied under the chin with a tape-string.

The nuns in their odd dresses, holding the babies in theirs, made Butterfly a great deal of amusement. At first, she kept close by Aunt Bessie, but gradually she became used to the sight, and soon was running around as freely from one baby to another as if she was going to learn how and stay as nurse. To her delight, she found the babies here just like other little human children, crowing and kicking, smiling and crying, eating and drinking and sleeping, all with their queer faces under the queer caps and above the queer dresses; "and queerest of all," said Butterfly, talking the matter over afterwards with Guy, "the beds with the dots of white pillows and the blue homespun bed-spreads, no bigger than Susy Millet's baby-house bed at home."

Butterfly would like to have spent the rest of the day here. She whispered the request to Aunt Matilda, but her aunt said "that the nunnery was only open for inspection an hour each day, that the end of the hour had come, and they must go;" so Butterfly began a series of kisses, taking the first baby, and going down the whole left-hand row. I think she would have done the same on the right, if the clock had not finished striking, and Guy's mother, taking out her pocket-book, deposited a present on the box placed upon a table for contributors' use. This said to Butterfly, "Now it is full time to go," and Guy informed her her kisses would keep until the next time.

There were many other places to be seen in Montreal; Hal to be visited in his own home, and Margaret in hers; excursions by land and excursions by water. Every day and hour of the time was used up to its last moment, and in the course of all many pleasant things occurred of which I should like to tell my readers, but I have not room. I must pass over what remained to be seen, and come to the last morning, when so much was to be done that even quiet and prompt Aunt Matilda began to feel that there was some doubt whether they should be ready in time for the cars. Hal and Margaret must be seen and told " good-by ;" and then, you know,

it took so long to make all those little
arrangements by which they should
"never, never, *never* forget each other;"
Guy and Hal must arrange a meeting at
Guy's home for the next summer; and
Guy must plan that Bertie and Bennie
should come at the same time. So Guy
said :

"We shall have a jolly time, I tell you
what, sir, with Tiger and Sprite, and the
ocean. Boys couldn't help it. Come on
and see !"

Margaret had brought a bunch of
English forget-me-nots with their loving
blue eyes to Butterfly, and had let the
flowers speak the words first for her; and
Butterfly had given Margaret one of her
golden curls, tied together with a blue
ribbon, and laid on a sheet of white paper,

with "Forever" written in her best handwriting in the circle within the curl.
Then she had kissed it as she gave it to
her. Margaret had kissed the same spot
afterwards, and so the little girls were
never to forget, and to be friends forever.

This was the last of Butterfly's sight-
seeing for the present. For one day and
one day only she would travel with Guy
on her way to Frostland, and then—
Butterfly felt the tears coming into her
eyes every time she thought of the separation. I dare say she would have
minded leaving Montreal and Hal and
Margaret a great deal more if this part-
ing had not been right before her.

When Montreal was left behind and
they were coming back to the United
States as fast as the cars could carry

them, Butterfly could think or speak of
nothing but her misery at leaving Guy;
and as for Guy, though he tried to whistle
and hum, buy papers and read a line here
and there, buy candy, and pears, and very
green apples, nothing would suffice. Big
boy and manly as he was, leaving Butterfly, not to see her on the next morning, or
the next, or the next, was too much for
him. Every now and then something
came between him and the landscape,
something between him and the pretty
face he had learned to love so well, something that needed brushing out of his eyes,
and of which he felt very much ashamed,
only all the shame did not help him in the
least.

There had never been a ride when he
and Butterfly said so little. I am sure it

was well it was no longer, for every mile only added to their trouble; and when the conductor called out at last, " Frostland station!" Guy was glad of the excuse given him by the bustle of departure of hiding his face away behind the pile of bags and shawls which he politely carried out for the aunts upon the platform; and when the bell rang, and the whistle sounded, and he knew he must go, he had no choice but to leap back again, call out, "Remember next summer at the beach!" and wave his hat as long as the figure of the little girl, with her blue eyes and her sweet, loving face, was to be seen.

As for Butterfly, I think her mother was surprised and grieved, after her long separation from her child, to have cheeks all wet with tears to kiss, and to hear among the very first things:

"Oh, mamma, if you please, we are all going to the seaside together next summer. May we, say, please, may we go?"